The Rise of E-Skin: Bridging the Gap Between Humans and Technology

Navya

TABLE OF CONTENTS

C h a p t e r   1

1.1 E-skin

Skin is the largest organ of the human body. It protects the body as a physical barrier and harbors sophisticated sensations of touch and heat, enabling us to perceive temperature changes, pressure and vibrations, texture and shapes of our surroundings. Mimicking the properties of human skin, wearable and flexible electronic devices provide an intimate yet noninvasive contact with human body and are capable of digital monitoring of physical and biochemical signals. Compared with bulky, wired clinical equipment, these miniaturized on-skin sensors are featured by real-time diagnosis and continuous monitoring, which open up new opportunities for long-term health assessment and disease diagnosis.

E-skin refers to integrated electronics that mimic and surpass the functionalities of human skin. Due to their flexible and conformable nature, e-skins may be placed on various robotic and human bodily locations for continuous biosignal monitoring, rivaling bulky medical equipment in the fields of robotics and prosthetics[1,2]. Engineered for self-contained operational frameworks, e-skins act as human-machine interfaces for smart bandages[3], wristbands[4], tattoo-like stickers[1], textiles[5], rings[6], face masks[7], as well as customized smart socks and shoes[8] for various applications. The integration of wearable sensors with the human body demands that the mechanical properties of the sensors should be soft, stretchable, and compliant to curved skin (**Fig. 1-1**). Compared with conventional rigid devices, soft e-skin patches seamlessly interface with the skin, achieving a conformal and stable contact that minimizes motion-induced artifacts and wearing discomfort[9]. The convenience and flexibility of applying these electronic patches to any target location, while continuously and noninvasively measuring multiplexed signals via mobile connectivity, has surpassed conventional point-of-care to become an ideal form of wearable systems. With the increasing demands for remote and at-home care, e-skins have been applied for personal fitness[4,10], VR[11,12], telemedicine and early disease detection[13,14], as well as COVID-19 tracing and monitoring[15,16].

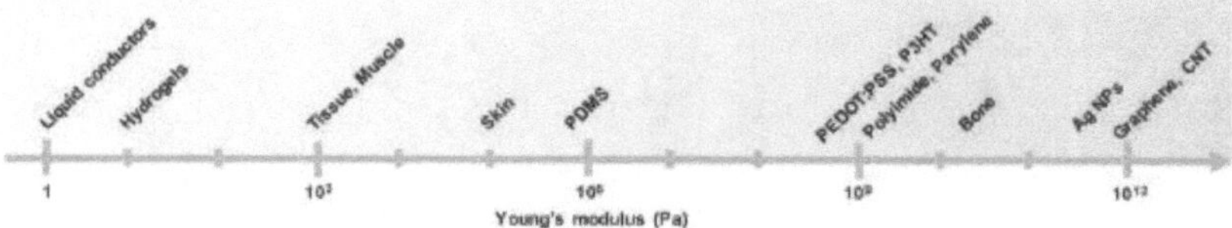

Figure 1-1. Soft electronic materials for skin-interfaced electronics. Comparison of Young's moduli among representative materials.

1.2 Emerging sensors landscape in e-skins for data acquisition

Training of an intelligent ML system requires a substantial amount of high-quality data. Unlike conventional clinical laboratory tests that are performed discretely and infrequently, emerging wearable sensors provide the ability for continuous acquisition of digitalized data with multiplexed sensors, allowing for more personalized care by analyzing deviations in individual baselines[17]. This approach greatly mitigates the biases from environmental factors such as diet, age, stress, and drug use, yielding a more appropriate and accurate medical diagnostic tool based on the individual rather than population-level statistics. Here we focus on the two primary sensing domains in e-skin platforms (**Fig. 1-2**), namely physical and biochemical sensors, highlighting their key usage and applications.

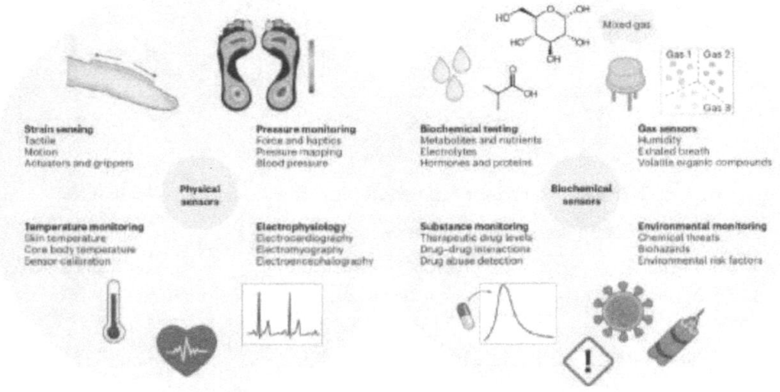

Figure 1-2. Emerging sensors in e-skin for health monitoring and robotics. The combination of physical and biochemical sensors provides access to force sensing and mapping, electrophysiology, as well as biochemical substances in body fluids and surroundings.

1.2.1 Strain and pressure sensing

A commonly integrated sensor, strain sensors track the resistance of electronic materials under deformations. These sensors enable the detection of large distortions from bodily motions[18] and small deviations for tactile perception[19]. As another motion sensing mechanism, pressure sensors utilize piezoresistive materials or capacitors with a pressure cavity. Similar to strain sensors, pressure sensors could be customized to perform pressure mapping[20,21], user interactive visualization[22,23], as well as tactile sensing[24,25].

To fully mimic skin sensations, strain and pressure sensors are often combined for haptic interfaces in HMI applications[11]. When placed near arteries, strain and pressure sensors can detect vital signs such as blood pressure and heart rate variability[26]. Recent studies have also utilized piezoelectric sensor arrays, which capture acoustic vibrations from tissue for blood pressure monitoring and imaging applications[27–29].

1.2.2 Temperature monitoring

While elevated core body temperatures often result from infections and overheating, a decreased temperature can lead to faltered physiological systems and even organ failure. Although e-skin sensors are commonly applied to monitor skin surface temperature, arrays of sensors could be used in conjunction to minimize local deviations and display an accurate temperature profile[30]. Further studies have investigated correlating skin surface temperatures to core body profiles[31]. In addition, temperature data is of significance for calibrating biochemical sensors, as chemical reactions are sensitive to their operating temperature[32].

1.2.3 Electrophysiology

Electrophysiology refers to measuring the electrical activities of tissues and organs. Common skin-interfaced biopotential modalities involve ECG[33], EMG[34,35], and EEG[36,37]. These signals are measured by placing arrays of electrodes on the skin at different locations. E-skin-based electrophysiology sensors commonly show high performance due to the conformal contact between the soft e-skin and body with a low contact impedance.

1.2.4 Biochemical sensing

E-skin-based biochemical sensors have been widely applied to analyze molecular biomarkers (e.g., electrolytes[38], metabolites[4], amino acids[10], neurotransmitters[39], and proteins[40]) in human biofluids including sweat[4,10,13,41], saliva[42], and interstitial fluids[43]. Common biosensing signal transduction strategies include electrochemical and optical detection mechanisms[44]. These sensors can be applied for a wide range of biomedical applications including fitness tracking, metabolic monitoring[4], cystic fibrosis diagnosis[38], gout management[13], and stress assessment[45].

1.2.5 Substance monitoring

In addition to natural biofluid components, e-skins can also detect substances that are extrinsic to the normal metabolism such as drugs[46] (e.g., vancomycin[47] and levodopa[48,49]), alcohol[50,51], caffeine[52], and heavy metals[53]. By focusing on personalized pharmacokinetics instead of population studies, continuous therapeutic drug monitoring can improve treatment outcomes and reduce side effects through dosage adjustments, which are especially important for drugs with narrow therapeutic windows[46]. Moreover, e-skin sensors can serve as a rapid screening tool for drug abuse[54,55].

1.2.6 Gas sensors

Human breath contains rich molecular information and could provide a noninvasive health profile like biofluids. Many VOCs in the breath are diagnostic biomarkers for infectious, metabolic, and genetic diseases[56,57]; for example, breath carbon monoxide is linked to neonatal jaundice and breath ammonia and nitric oxide are connected to asthma[58]. Integrated sensor arrays known as electronic noses have been developed to detect humidity, VOCs and other gas components in exhaled breath and the surrounding environment[59]. Combined with ML, these sensors can distinguish complex chemical signatures[60,61], and have been employed for breath-based individual authentication[62], soil nitrogen assessment[63], and evaluating food freshness[64].

1.2.7 Environmental monitoring

Environmental risk factors, including chemical threats and pathogenic biohazards, pose a risk to both the human body and safe robotic operations. AI-powered e-skins have expanded their scope to encompass not only monitoring the human body but also the surrounding environment. During remote operations, e-skin systems can detect trace amounts of dangerous compounds and provide environmental feedback without human exposure[2]. A combination of biochemical sensors was integrated into an e-skin patch attached to a robotic arm that could detect hazardous materials including nitroaromatic explosives, pesticides, nerve agents, and infectious pathogens with autonomous ML-based decision-making algorithms[2].

1.3 ML pipeline for data analysis

While emerging e-skin is revolutionizing robotics and medical practices by continuously monitoring multimodal data[65], data analysis is playing an increasingly important role for interpreting the large, complex biological profiles generated from various sensors. Conventional analysis of e-skin data largely relies on human supervision, where signal processing and data evaluation is time-consuming and interpreted from a restricted point of view[1,4,5]. There is an unmet demand between e-skin hardware and efficient data analysis solutions. Recent developments in deep learning have permitted the evaluation and even generation of big data for health applications[66]. AI can reveal medical insights that are challenging to acquire with traditional data-analytics while providing accurate predictions that can mimic or even surpass human expertise[67–69]. AI together with the rapidly growing interest in health monitoring and remote robotics have become the main catalyst pushing forward advanced e-skin innovations.

In a typical ML pipeline (**Fig. 1-3**), raw data collected from e-skins will first be preprocessed for feature extraction. Popular preprocessing techniques include filtering, smoothing, downsampling with a sliding window, dimensionality reduction, as well as baseline removal and normalization[70]. An ML algorithm is then selected for the specific objective, which can be supervised or unsupervised, classification or regression, discriminative or generative[70]. During model selection, one needs to account for data availability[68]. While simple models may struggle to represent the expected trends, complex models on simple datasets may lead to non-reproducible conclusions, particularly in health applications when a small dataset may be specific to a particular demographic.

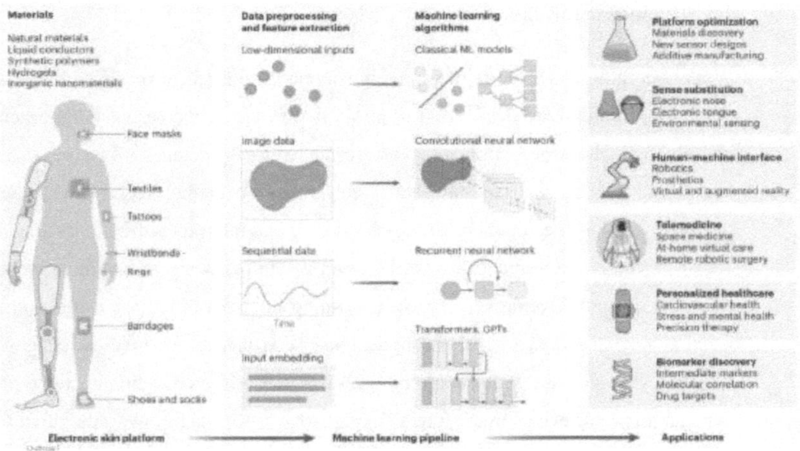

Figure 1-3. Overview of AI-powered e-skin and ML pipelines. E-skin provides access to human information or serves as an interface to robotics by continuous and noninvasive monitoring of multimodal physical and biochemical sensors. The data stream is constructed and transformed into a standard numerical format through data preprocessing and feature extraction. Based on the intrinsic data properties, different ML algorithms can be selected and trained, allowing for real-world applications.

1.4 Outline of the book

This book consists of two parts. Here we briefly summarize these two parts and their organization:

Part 1: Fundamental Aspects of Wearable Sensors in E-Skin (Chapters 1-4

Part 1 consists of Chapter 1 to 4, where the effort is focused on fundamental side of wearable sensors in e-skin. Chapter 2 details the development of multimodal physiological and biochemical sensors that enable long-term continuous monitoring with high sensitivity and stability, with an emphasis on our strategy of significant sensor performance improvement and mechanism analysis. Chapter 3 and 4 explore the application of integrating these multimodal and robust sensors in robotics and wound monitoring and treatment.

Part 2: ML-Assisted Data Analysis and Future Prospects (Chapters 5-6)

Part 2 consists of Chapter 5 and 6, where the ML-assisted data analysis is implemented in addition to the multimodal sensors. Chapter 5 highlights the transformative deployment of AI in deconvoluting multimodal data analysis and unveiling health profiles on mental health. Chapter 6 discusses challenges and prospects of AI-powered e-skins, offering predictions for the evolution of smart e-skins.

WEARABLE MULTIMODAL SENSORS IN E-SKIN

2.1 Integrated physiological sensors with high sensitivity

The CARES platform contains multiple physical sensors to monitor stress-related vital signs. We placed a capacitive pressure sensor above the radial artery for pulse waveform monitoring (**Fig. 2-1a-e**). Because of the soft PDMS-engraved airgap, the pressure sensor is highly sensitive to soft pressure loads (such as a feather), with an impressive sensitivity of 113.1% kPa^{-1} under the range of 0–500 Pa. The pressure sensor also displays highly robust performance and mechanical stability during a repetitive pressure-loading test involving 5,000 cycles, mimicking daily use on the skin (**Fig. A-1**). A printed resistive temperature sensor was integrated into the CARES for skin temperature recording *in situ* with a sensitivity around 0.115% °C^{-1} in physiological temperature ranges between 25–50 °C (**Fig. 2-1f** and **Fig. A-2**). Considering that temperature has a strong influence on enzymatic activities, the temperature information is used for calibrating the response of the three enzymatic biosensors to achieve highly accurate *in situ* metabolic analysis (**Figs. A-3** and **4**). It should be noted that other environmental factors such as humidity showed minimal influence on the performance of our chemical sensors (**Fig. A-5**). Additionally, a pair of printed Ag electrodes were used as a GSR sensor which demonstrated high conductivity compared with commercial gel electrodes (**Fig. 2-1g**). Owing to the ultrathin flexible polyimide substrate and strong interfacial strength enabled by the medical adhesive, the CARES showed excellent skin contact and mechanical resilience against undesirable physical deformations during continuous operations (**Figs. A-6** and **7**). The impermeable polyimide packaging also eliminated the influence of humidity from environmental surroundings and sweat (**Fig. 2-1h**).

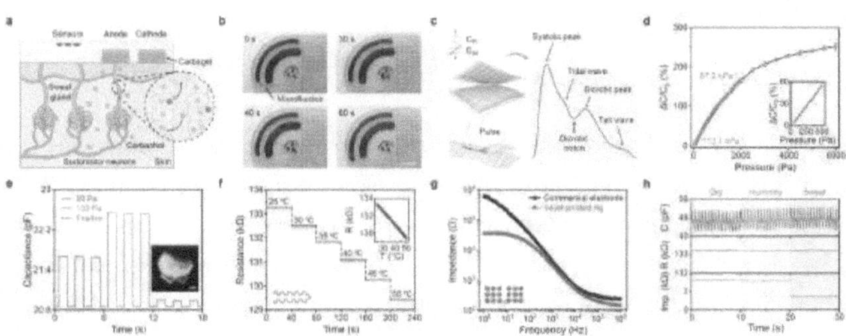

Figure 2-1. Design and characterization of highly robust physical sensors. a,b, Schematic (**a**) and on-body evaluation (**b**) of the microfluidic iontophoresis module for autonomous sweat induction and sampling at rest. Timestamps in **a** represent the period after a 5-min iontophoresis session. **c,** Schematic of the pressure sensor and a pulse waveform measured at the wrist. PI, polyimide. **d,** Pressure versus capacitance (C) characterizations of the pressure sensor. C_0, flat-state C. **e,** Repetitive response of the pressure sensor upon small pressure loads. Inset, a goose feather placed on a sensor. Scale bar, 1 cm. **f,** Response of the temperature (T) sensor in the physiological temperature range. R, resistance. **g,** Impedance of the skin/electrode interface measured with inkjet-printed Ag electrodes and commercial electrodes for GSR monitoring. **h,** Performance of encapsulated pulse, T and GSR sensors under environmental humidity and body sweat test. All error bars represent the SD from three sensors.

For *in vitro* temperature and GSR sensor characterizations, an amperometric method was used with an applied voltage of 1 V using a dual-channel electrochemical workstation (CHI 760E).

For *in vitro* pulse sensor characterizations, a parameter analyzer (Keithley 4200A-SCS) was applied to record the fast-changing capacitive signals at a sampling frequency of around 137 Hz. The influence of mechanical deformation on the physical sensor performance was investigated through pressing-releasing for 5,000 cycles using a Mark-10 force gauge. The influence of humidity was investigated by immersing the subject hand with the CARES in a customized glove box with a humidity gauge.

2.2 Biosensor mechanisms

2.2.1 Electrochemical potentiometric sensors

Wearable potentiometric sensors measure the passive open circuit potential between the working electrode and reference electrode (**Fig. 2-2a**). Such electrochemical sensors consist of a sensing electrode that is modified with a target-sensitive component and a reference electrode that maintains a stable potential in different solutions such as sweat. Most wearable potentiometric sensors are based on ISEs, which use a non-destructive measurement that converts ionic signals to electric potentials. The measured potential signal follows the well-known Nernst equation[1]

$$E = E_0 + \frac{RT}{nF} \ln a_I$$

where E is cell potential, E_0 is standard potential when $a_I = 1$, R is the universal gas constant, T is absolute temperature, n is the number of electrons involved, F is the Faraday constant, a_I is ionic activity quotient in aqueous and solid membrane phases. From the Nernst equation, the theoretical potential response limit of ISEs is around 59.2 mV at 25 °C per decade change of target analyte concentration. Due to its low-cost, simple operations, and reliable and continuous measurement, ISEs have been widely applied in detection of trace-level metals,[2,3] drugs,[4,5] and organics.[6] Conventional ISEs are composed of liquid contact, which is unfavorable for miniaturized designs and wearable usage. Solid-contact ISEs, on the other hand, have enabled the continuous detection of ions such as Na^+, K^+, H^+, Ca^{2+}, using potentiometric sensors in biofluids,[7–11] as well as heavy metal pollutants.[12]

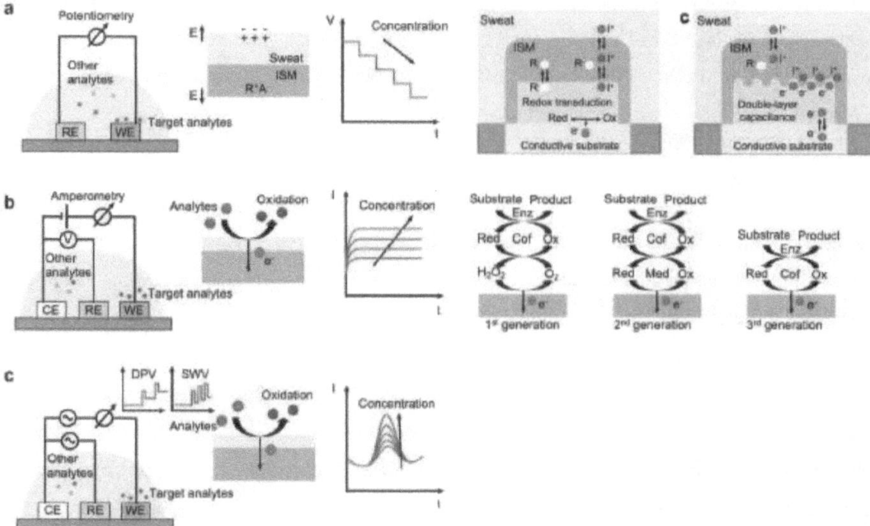

Figure 2-2. Sensing mechanisms of electrochemical biosensors. a, Schematic of potentiometry operating mechanism, sensor configurations, and ion-to-electron transducers using redox electroactive materials as well as nanomaterials. **b,** Schematic of amperometry operating mechanism,

sensor configurations, and three generations of amperometric enzymatic sensors. **c**, Schematic of differential pulse voltammetry/square wave voltammetry operating mechanism and sensor configurations.

The typical structure of an ISE consists of a sandwiched structure including an electron-conductive electrode substrate, an ion-to-electron transduction layer, and an ISM layer. A variety of solid-contact ISE materials have been studied as ion-to-electron transducers, mainly into two categories:[13] redox capacitance using electroactive materials, and double-layer capacitance using nanomaterials. ISMs typically contain an ionophore, lipophilic anionic sites, and a polymer matrix with its plasticizer. The polymer matrix is ion-permeable and enables strong mechanical stability to protect the ionophore and anionic sites from leaching and damage during operation. While primary ions could migrate into and out of the ISM, the ionophore binds to specific target ions across the solution/membrane interface. Due to such permselectivity, only the target ions can be transduced to the electrode substrate. Lipophilic anionic sites, also known as ion-exchangers, facilitate a charge neutral state of the ISM in order to attract ion movement. Meanwhile, the reference electrode is usually based on Ag/AgCl with saturated Cl^- ions in a polymer matrix to keep a stable reference potential.[14]

2.2.2 Electrochemical amperometric sensors

Wearable amperometric sensors measure current signals of redox reactions during the target substrate to product process at a constant applied potential. Such electrochemical sensors typically consist of a 3-electrode configuration, in which a working electrode is immobilized with a target-sensitive component, a counter electrode that forms a closed circuitry, and a reference electrode with a stable potential (**Fig. 2-2b**). The most common amperometric sensors are enzymatic sensors, in which the highly target selective and sensitive enzymes are immobilized onto the electrode surface in a hydrophilic porous matrix. For an ideal amperometric sensor, the current density is linearly proportional to the target concentration, allowing for immediate quantification for analytes.

Enzymatic sensors consist of three generations depending on the direct/indirect detection and electron transfer mechanism. A 1st generation enzymatic sensor is developed by monitoring hydrogen peroxide (H_2O_2) in the enzymatic reaction at the electrode surface. Most enzymatic sensors utilize oxidases, for instance, GOx complexed with coenzyme flavin adenine dinucleotide (FAD), or glucose dehydrogenase (GDH), in conjunction with cofactors (e.g., flavin adenine dinucleotide (FAD) and pyrroloquinoline quinone (PQQ)) are widely adopted for enzymatic glucose sensors.[15] GOx oxidizes glucose into gluconic acid in the presence of water and oxygen, and H_2O_2 is further oxidized, during which electron transfer is accomplished through cofactor FAD. 1st generation enzymatic sensing is an indirect monitoring method and the oxidase reaction relies on the use of oxygen as an electron acceptor. Once the oxygen concentration fluctuates in the solution, error readings can occur.

To eliminate oxygen dependence, 2nd generation enzymatic sensors replace oxygen with a synthesized mediator as the electron acceptor, shuttling electrons from GOx to the electrode. A representative mediator is PB, which is widely adopted due to its high electrocatalyst of H_2O_2 reduction in the presence of O_2 relative to conventional electrodes (e.g., platinum) and its low redox potential at around 0 V.[16] The 3rd generation enzymatic sensors are mediator-free and enable direct electron transfer between enzyme and electrodes, which reduces the influence of mediator deterioration. This is usually realized by using nanomaterials such as CNTs[17,18] and MOFs.[19]

2.2.3 Electrochemical direct oxidation

In addition to chronopotentiometry and chronoamperometry, a number of electrochemical techniques, including square wave voltammetry,[20] differential pulse voltammetry,[21] fast-scan CV,[22] and LSV,[23] have been developed to realize direct target detection and bypass the need of target-sensitive components such as ionophores and enzymes. Square wave voltammetry and differential pulse voltammetry are among the most common approaches for detecting electroactive materials. Similar to amperometric configurations, differential pulse voltammetry/square wave voltammetry based sensing involves a 3-electrode system, including a working electrode with high material stability and high surface area, a counter electrode that forms a closed circuitry, and a reference electrode with a stable potential (**Fig. 2-2c**). A time-dependent staircase excitation waveform of pulsed step voltammetry was applied between the working and counter electrodes to increase the ratio between the faradaic and nonfaradaic currents, and the current response is recorded before and at the end of the pulsed voltage. Reduction occurs at the working electrode surface when the potential is lower than the redox potential, and at potentials higher than the redox potential, oxidation occurs.[24]

During redox reactions, electroactive materials lose or gain electrons accordingly, resulting in current response with a limit of detection down to nanomols.[25]

2.3 Biochemical sensors for long-term wound monitoring

2.3.1 Materials and methodology for sensor fabrication

Tetrahydrofuran, dimethylformamide, PVB, sodium chloride, ammonium chloride, gelatin (from bovine skin), sodium thiosulfate pentahydrate, sodium bisulfite, ammonium ionophore I, aniline, chondroitin 4-sulfate, 1,4-Butanediol diglycidyl ether, EDOT, NaPSS, PU, GOx, uricase, chitosan, iron (III) chloride, potassium ferricyanide (III), paraformaldehyde, UA, multi-walled CNTs were obtained from Sigma Aldrich. LOx was purchased from Toyobo Corp. Hydrochloric acid, acetic acid, methanol, ethanol, acetone, urea, and dextrose (D-glucose), and Dulbecco's PBS were acquired from Fisher Scientific. The SEBS polymer was obtained from Asahi Kasei Corporation. TCP-25 (GKYGFYTHVFRLKKWIQKVIDQFGE) (98% purity, acetate salt) and tetramethylrhodamine labeled TCP-25 were purchased from CPC Scientific.

Enzymatic sensors preparation. To increase the electrode surface area for enzymatic sensors, a nanostructured Au film were electrodeposited on Au electrodes in a solution containing 50 mM chloroauric acid and 0.1 M HCl using multi-potential deposition for 1500 cycles (for each cycle, -0.9 V for 0.02 s and 0.9 V for 0.02 s). For glucose and lactate sensors, a PB layer was deposited onto the Au electrodes by 10 cycles of CV (-0.2–0.6 V vs. Ag/AgCl) with a scan rate of $50\,mV\,s^{-1}$ in a freshly made solution containing 2.5 mM $FeCl_3$, 2.5 mM $K_3[Fe\,(CN)_6]$, 100 mM KCl, and 100 mM HCl. For the UA sensor, a PB layer was deposited using the same approach except only one CV cycle of electrodeposition. Next, a chitosan solution was prepared by dissolving 1% chitosan in a 2% acetic acid solution followed by vigorous magnetic stirring for 1 h. The resulting solution was then mixed with CNTs ($2\,mg\,ml^{-1}$) by ultrasonic agitation over 30 min to prepare a chitosan/CNTs solution. To prepare all enzymatic sensors, the chitosan/CNTs solution was mixed thoroughly with an enzyme solution ($10\,mg\,ml^{-1}$ in PBS, pH 7.2) with a volume ratio of 2:1. Next, 1 µl of the enzyme/chitosan/CNTs cocktail was drop-casted onto the PB/Au electrode and dried under 4 °C. Finally, the PU layer was prepared by drop-casting 4.5 µl 15 mg ml⁻¹ PU solution in a solvent mixture containing tetrahydrofuran and dimethylformamide (volume ratio 98:2) on the enzyme layer and air-dried overnight under 4 °C.

pH sensor preparation. The pH sensor was based on pH-sensitive polyaniline film deposited on a Au electrode. First, the working electrode was electrochemically cleaned *via* 10 cycles of CV with a scan rate of 0.1 V s^{-1} in 0.5 M HCl (-0.1–0.9 V). Next, the polyaniline electro-polymerization was performed in a 50 µl solution containing 0.1 M aniline and 1 M HCl *via* 12 CV cycles (-0.2–1.0 V) with a scan rate of 0.1 V s^{-1}. The fresh solution was then used for another 12 CV cycles. Finally, pH electrodes were air-dried overnight.

Ammonium sensor preparation. A PEDOT:PSS film was electrodeposited using constant current of 0.2 mA cm^{-2} for 10 min in a solution prepared by dissolving ferrocyanide (30 mg), NaPSS (206.1 mg), EDOT (10.7 µl) in 10 ml DI water. Next, an NH_4^+ selective membrane cocktail solution was prepared by dissolving 1 mg ammonium ionophore I, 33 mg PVC, and 66 mg DOS in 660 µl tetrahydrofuran. A 1.5 µl of the cocktail solution was then drop-casted on the PEDOT layer to create an ammonium selective membrane and air-dried overnight.

Reference electrode preparation. To prepare the Ag/AgCl reference electrode, silver was electrodeposited at −0.2 mA for 100 s using a plating solution containing 250 mM silver nitrate, 750 mM sodium thiosulfate and 500 mM sodium bisulfite. 10 µl solution of 0.1 M $FeCl_3$ was dropped on the Ag electrode for 90 s. Next, a solid-state reference membrane cocktail was prepared by dissolving 78.1 mg of PVB and 50 mg of NaCl in 1 ml of methanol followed by vigorous agitation in an ultrasonic bath for 30 minutes. Next, a 2.5 µl of the reference cocktail membrane was drop-casted on the Ag/AgCl electrode surface and air-dried overnight.

In vitro sensor characterization: The multiplex sensor patches were characterized to evaluate their sensitivity, stability, and reproducibility in solutions of target analytes in simulated wound fluid using a 1000C Multi-Potentiostat (8-channel) (CH Instruments, Inc., Austin, TX, USA). The simulated wound fluid solution was prepared by dissolving 584.4 mg NaCl, 336.0 mg $NaHCO_3$, 29.8 mg KCl, 27.8 mg $CaCl_2$, and 3.30 g BSA in 100 mL DI water. The enzymatic sensors were characterized chronoamperometrically in 0–40 mM glucose, 0–4 mM lactate, and 0–150 µM UA, at a potential of 0 V. The pH sensor calibration was performed in McIlvaine buffer solutions. Both pH and ammonium sensors were characterized electrochemically using open circuit potential.

2.3.2 Results and discussion

The array of flexible biosensors was custom developed to allow real-time multiplexed monitoring of the biomarkers in complex wound exudate. The continuous and selective measurement of glucose, lactate, and UA is based on amperometric enzymatic electrodes with GOx, LOx, and uricase immobilized in a highly permeable, adhesive, and biocompatible chitosan film, respectively (**Fig. 2-3A**). Electrodeposited PB serves as the electron-transfer redox mediator for the enzymatic reaction which allows the biosensors to operate at a low potential (~0.0 V) to minimize the interferences of oxygen and other electroactive molecules. Due to the complex and heterogeneous composition of wound fluid (e.g., high protein levels, local and migrated cells, and exogenous factors such as bacteria) [26], previously reported enzymatic sensors suffer from severe matrix effects and fail to accurately measure the target metabolite levels in untreated wound fluid (**Figs. A-8, A-9** and **Note A-1**). Moreover, high levels of metabolites in diabetic wound fluid, especially glucose (up to 50 mM), pose another major challenge to obtain linear sensor response in the physiological concentration ranges. To address these issues and achieve accurate wound fluid metabolic monitoring, increase sensor range, and minimize biofouling effects, we explored the use of an outer porous membrane that serves as a diffusion limiting layer to protect the enzyme, tune response, increase operational stability, as well as enhance the linearity and sensitivity magnitude of the sensor. We fabricated our enzymatic GOx/chitosan/MWCNTs glucose sensor with additional porous membrane coatings including chitosan, poly(ethylene glycol) diglycidyl ether (PEGDGE), Nafion, and PU (**Fig. A-9**). As expected, the addition of diffusion layers indeed improves the sensor's linear range in simulated wound fluid. However, chitosan, PEGDGE, and Nafion coated sensors did not show reliable responses in wound fluid upon the addition of glucose. The PU-based enzymatic sensors showed the highest linearity over the wide physiological concentration range as well as high reproducibility in complex wound fluid matrix (**Fig. A-10**). The amperometric current signals generated from the PU-coated enzymatic glucose, lactate, and UA sensors are proportional to the physiologically relevant concentrations of the corresponding metabolites in simulated wound fluid with sensitivities of 16.34, 41.44, and 189.60 nA mM^{-1}, respectively (**Fig. 2-3B–D**). Continuous monitoring of ammonium is based on a potentiometric

ISE where the binding of ammonium with its ionophore results in an electrode potential log-linearly corresponding to the target ion concentration with a sensitivity of 59.7 mV decade^{-1} (**Fig. 2-3E** and **F**). Similarly, the pH sensor utilizes an electrodeposited polyaniline film as the pH-sensitive membrane and shows a sensitivity of 59.7 mV per pH (**Fig. 2-3G**). For all chemical sensors, a PVB-coated Ag/AgCl electrode was used as the reference electrode which provides a stable voltage independent of the variations of wound fluid compositions [27]. A gold microwire-based resistive temperature sensor is integrated as part of the sensor array and shows a sensitivity of approximately 0.21% °C^{-1} in the physiological temperature range of 25–45 °C (**Fig. 2-3H**).

Considering that other electrolytes and metabolites present in wound fluid may negatively affect the sensor outputs, we examined the selectivity of the sensor array consisting of all six sensors. As illustrated in **Fig. 2-3I**, the addition of non-target electrolytes and metabolites did not trigger any substantial interference to the sensor response. Moreover, all biosensors showed high selectivity over non-specific compounds when evaluated in simulated wound fluid (**Fig. A-11**). It should be noted that while temperature has negligible effects on the potentiometric sensors, it significantly influences the performance of the enzymatic sensors due to the temperature-dependent enzyme activities (**Fig. A-12**). Moreover, our data show that the medium pH could also impact the performance of enzymatic sensors (**Fig. A-13**). With pH and temperature sensors integrated into the wearable patch, we are able to perform real-time adjustments and calibration of the enzymatic biosensors based on temperature and pH variations to realize accurate wound metabolite analysis.

Owing to the soft SEBS substrate and the serpentine-like design of electronic interconnects, the wound patch showed excellent mechanical flexibility and stretchability, which are essential to maintaining good contact with the skin *in vivo* during the chronic wound healing process. Negligible alterations in the sensor responses before and under unidirectional tensile stretching (**Fig. 2-3I**) and after repetitive mechanical bending (**Fig. A-14**) were observed, indicating highly consistent sensor performance under various physical deformations.

As the sensor patch is designed for long-term *in vivo* use, its cytocompatibility and biocompatibility are of great importance. Cell viability and metabolic activity of the cells seeded on a multiplexed sensor array were analyzed using a commercial live/dead kit, and PrestoBlue assay, respectively (**Fig. 2-3K–N** and **Fig. A-15**). The high cell viabilities shown in the representative live/dead staining images of human dermal fibroblasts (HDF) and normal human epidermal keratinocytes (NHEK) cells (**Fig. 2-3K–M** and **Fig. A-15**) along with the consistently increased cell metabolic activities (**Fig. 2-3N**) over multi-day culture periods, indicate the high cytocompatibility of the soft sensor patch.

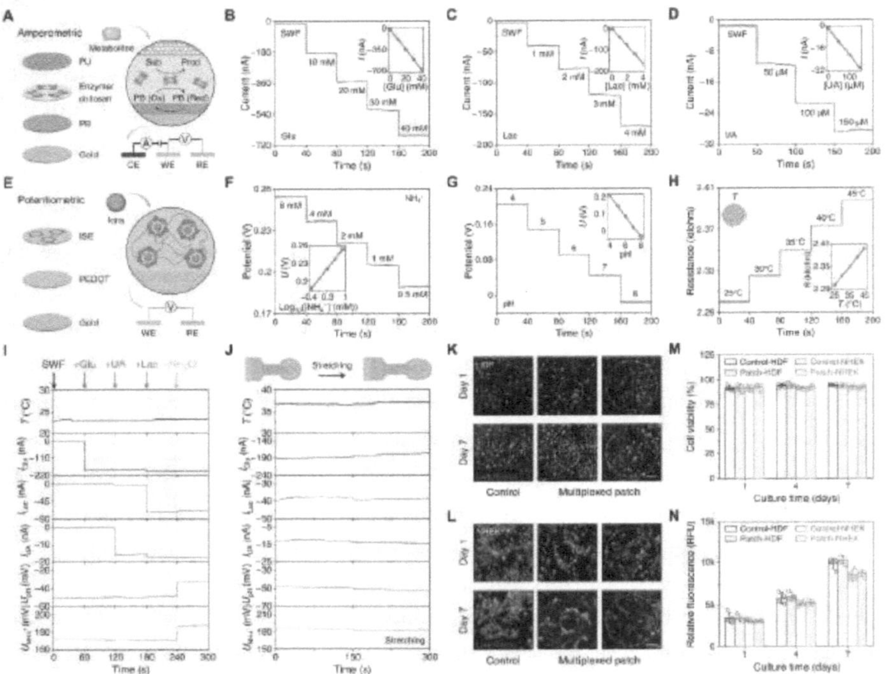

Figure 2-3. Design and characterization of the sensor array for multiplexed wound analysis. (A to D) Schematic (**A**) and chronoamperometric responses of the enzymatic glucose (**B**), lactate (**C**), and UA (**D**) sensors in simulated wound fluid. Insets in **B–D**, the calibration plots with a linear fit. Sub, substrate; Prod, product; CE, counter electrode; WE, working electrode; RE, reference electrode; I, current. (**E and F**) Schematic (**E**) and potentiometric response (**F**) of an NH_4^+ sensor in simulated wound fluid. Insets in **F**, the calibration plot with a linear fit. U, potential. (**G**) Potentiometric response of a polyaniline-based pH sensor in McIlvaine buffer. Insets, the calibration plot with a linear fit. (**H**) Resistive response of an Au-microwire-based temperature sensor under

temperature changes in physiologically relevant range in simulated wound fluid. Insets, schematic of a temperature sensor and the calibration plot with a linear fit. All error bars in **A–H** represent the standard deviation (s.d.) from three sensors. (**I**) Selectivity study of the multiplexed sensor array in simulated wound fluid. 10 mM glucose, 50 μM UA, 1 mM lactate, and 1 mM NH_4^+ was added sequentially to the simulated wound fluid. **J**, Responses of the multiplexed sensor array before and during mechanical stretching (15%) in simulated wound fluid (pH 8) containing 10 mM glucose, 50 μM UA, 1 mM lactate, and 0.25 mM NH_4^+. (**K and L**) Representative live (green)/dead (red) images of human dermal fibroblasts (HDF) (**K**) and normal human epidermal keratinocytes (NHEK) (**L**) cells seeded on the multiplexed sensor array and in PBS (control) after 1-day and 7-day culture. Scale bars, 200 μm. (**M and N**) Quantitative analysis of cell viability images (**M**) and cell metabolic activity (**N**) over a 7-day period post culture. Error bars represent the s.d. (n=4).

2.4 Biochemical sensors for long-term sweat monitoring

We introduce a general approach to prepare highly stable and sensitive electrochemical biosensors, which utilizes analogous composite materials for stabilizing and conserving sensor interfaces. The obtained biochemical sensors achieved a record-breaking long-term stability of more than 100 hours of continuous operation with minimal signal drifts (amperometric signals decaying less than 0.07% h^{-1} and potentiometric signals drift less than 0.04 mV h^{-1}), which greatly exceeds those obtained with previous widely adopted wearable sweat sensors.

2.4.1 Materials and methodology for sensor fabrication

The SEBS was provided by the Asahi Kasei Corporation. UA, sodium tetraphenylborate, and glutaraldehyde (25% aqueous solution) were purchased from Alfa Aesar. Agarose, carbachol, BSA, gold chloride trihydrate, hydrochloric acid, iron(III) chloride, potassium ferricyanide (III), potassium ferrocyanide (IV), PVB, PVC, DOS, EDOT, NaPSS, aniline, L-lactic acid, sodium ionophore X, sodium tetrakis[3,5-bis(trifluoromethyl) phenyl] borate, valinomycin, nonactin, tetrahydrofuran, toluene, GOx from Aspergillus niger (216 U mg^{-1}), uricase from Bacillus fastidiosus (15.6 U mg^{-1}) were purchased from Sigma-Aldrich. Methanol, ethanol, sodium chloride, potassium chloride, nickel chloride, urea, L-ascorbic acid, dextrose (D-glucose) anhydrous, PBS were purchased from Thermo Fisher Scientific. LOx (106 U mg^{-1}) was purchased from Toyobo Co. Medical tapes were purchased from 3M (468 MP). PET films (12 μm thick) were purchased from McMaster-Carr. Polyimide (PI-2611) was purchased from HD MicroSystems, Inc. PDMS (SYLGARD 184) was purchased from Dow Corning. Polyimide film (12.5 μm) was purchased from DuPont. STAI questionnaire license was purchased from Mind Garden, Inc.

Enzymatic sensor preparation: An electrochemical workstation (CHI 760E, CH Instruments, USA) was used to prepare enzymatic biosensors. Pulsed voltammetry from -0.9 V to 0.9 V (3000 cycles in total) in 50 mM $HAuCl_4$ was used to first deposit Au nanoparticles (AuNPs) on the carbon electrode at a signal frequency of 50 Hz, in order to increase surface area and enhance sensitivity. A thin PB transducer layer was deposited by applying cyclic voltammetry for 2 cycles for glucose and UA, and 4 cycles for lactate (from -0.2 V to 0.6 V with a scan rate of 50 mV s^{-1}) in a fresh solution consisting of 2.5 mM $FeCl_3$, 2.5 mM $K_3Fe(CN)_6$, 100 mM KCl and 100 mM HCl. The electrodes were then deposited with a NiHCF protection layer by applying cyclic voltammetry for 50 cycles (from 0V to 0.8 V with a scan rate of 100 mV s^{-1}) in a fresh solution containing 0.5 mM $NiCl_2$, 0.5 mM $K_3Fe(CN)_6$, 100 mM KCl and 100 mM HCl. The electrodes were then dried before drop-casting enzyme cocktail. For all three amperometric enzymatic sensors, the enzyme cocktails were prepared as follows: BSA (1% w/w), 2.5% glutaraldehyde (2% v/v), and 10 mg mL^{-1} enzyme (4% v/v) was mixed in 1 mL PBS. 0.5 μL enzyme cocktail was drop-casted onto each enzymatic sensor electrode surface and then dried at 4 °C overnight. For the lactate sensor, a limit diffusion membrane was further drop-casted by applying 0.5 μL solution containing 17 mg PVC and 65 mg DOS in 660 μL tetrahydrofuran.

Reference electrode preparation: To prepare the shared reference electrode, 10 μL of 0.1 M $FeCl_3$ solution was drop-casted onto the Ag surface for 20 s and rinsed with deionized water, and then 1.5 μL of PVB reference cocktail was applied on the Ag/AgCl surface by dissolving 79.1 mg PVB and 50 mg NaCl into 1 mL methanol and left drying overnight.

ISE sensors preparation: The Na^+ selective cocktail was prepared as follows: 1 mg of Na ionophore X, 0.55 mg sodium tetrakis[3,5-bis(trifluoromethyl) phenyl] borate, 30 mg PVC, 30 mg SEBS, and 65 mg DOS were dissolved in 660 μL tetrahydrofuran. The K^+ selective cocktail was prepared as follows: 2 mg of valinomycin, 0.5 mg sodium tetraphenylborate, 30 mg PVC, 25 mg SEBS, and 70 mg DOS were dissolved in 350 μL tetrahydrofuran. The NH_4^+ selective cocktail was prepared as follows: 1 mg of nonactin, 30 mg PVC, 30 mg SEBS, and 65 mg DOS were dissolved in 660 µL tetrahydrofuran. The inkjet carbon electrode was activated in 0.5 M HCl with cyclic voltammetry scans of 10 cycles (-0.1 V to 0.9 V with a scan rate of 100 mV s^{-1}). The electrodes were then baked in a vacuum oven at 120 °C for 1 hour to remove moisture. 2 µL of Na$^+$ selective cocktail, 2 µL of K$^+$ selective cocktail, and 2 µL of NH$_4^+$ selective cocktail was drop-casted onto the carbon electrode and dried overnight.

In vitro sensor characterization: To obtain the best performance for long-term continuous measurements, all sensors were placed in a buffered solution containing 100 µM glucose, 5 mM lactate, 25 µM UA, 40 mM NaCl, 8 mM KCl, 2 mM NH$_4$Cl for 30 minutes to minimize the potential drift. All the *in vitro* biosensor characterizations were performed with cyclic voltammetry and amperometric i-t through a multi-channel electrochemical workstation (CHI 1430, CH Instruments, USA). For in vitro enzymatic sensor characterizations, analyte solutions were prepared in PBS, with glucose ranging from 0−100 µM, lactate ranging from 0−20 mM, and UA ranging from 0−100 µM. For in vitro ISE sensor characterizations, analyte solutions were prepared in deionized water, with NaCl ranging from 10−160 mM, KCl ranging from 2−32 mM, NH$_4$Cl ranging from 0.5−8 mM. The enzymatic sensors were characterized chronoamperometrically at a potential of 0 V, and ISE sensors were characterized using open circuit potential measurement. Both potentiometric and chronoamperometric responses were set as 1 s sampling interval, except for long-term monitoring where the sampling interval was set as 10 s to minimize data overload. To test the pH influence on PB-NiHCF-based enzymatic biosensors, McIlvaine buffer solutions were prepared and calibrated containing 0−100 µM H$_2$O$_2$. Temperature influence characterizations were carried out on a ceramic hot plate (Thermo Fisher Scientific).

To characterize the stability of the PB and PB-NiHCF electrodes, dissolved Fe^{2+} concentrations were determined by ICP−MS using an Agilent 8800. The sample introduction system consisted of a micromist nebulizer, scott type spray chamber and fixed injector quartz torch. A guard electrode was used and the plasma was operated at 1500 W. All elements were measured in Helium MS/MS mode.

The morphology of materials was characterized by field-emission SEM (Nova 600). Cross-sectional lamella was prepared by standard focus ion beam cutting (FIB, Nova 600). The STEM characterizations and EDS analyses were performed using a JEOL JEM-ARM300CF S/STEM system (300 keV).

2.4.2 Results and discussion

A number of electrochemical sensing strategies based on enzymes[7], ionophores[10], molecularly imprinted polymers[23], aptamers[20], and antibodies[28] are reported, where the majority of existing wearable chemical sensors are primarily based on amperometric enzymatic sensors or potentiometric ISEs as these sensors could offer real-time continuous monitoring with high temporal resolution. However, one main bottleneck for the practical applications of these sensors is their limited operation lifetime and long-term stability during continuous wearable sensing. Large sensor drifts are evident when they are used in body fluids which substantially hinder the long-term continuous usability of wearable chemical sensors.

Most wearable enzymatic biosensors are based on PB, which serves as an efficient electron-transfer mediator with a low redox potential of around 0 V. However, PB-based biosensors suffer from poor stability during long-term use in biofluids because PB degrades in neutral and alkaline solutions as the hydroxide ions (OH^-), a product of H_2O_2 reduction, can break the Fe$-$(CN)$-$Fe bond (**Note A-2**). In order to stabilize PB while retaining its outstanding catalytic activity, we utilize a PB-analogue NiHCF with a similar zeolitic crystal structure that is catalytically inactive but forms a stabilized solid solution composite, protecting the PB sensor interface (**Fig. 2-4a**). Additionally, the enzymes were protected in a glutaraldehyde-crosslinked BSA matrix. To fabricate enzymatic sensors, AuNPs were first electrodeposited onto an inkjet-printed inert carbon electrode to possess a high electroactive area for sensitive electrochemical sensing followed by PB-NiHCF deposition. STEM and EDS analyses (**Fig. 2-4b** and **Fig. A-16**) indicate that NiHCF forms a thin protective layer on PB with an obscure boundary. Our electrochemical characterizations confirmed that, compared to PB which suffered from rapid degradation during electrochemical measurement and other transition metal hexacyanoferrates (i.e., PB-CoHCF and PB-CuHCF), PB-NiHCF could withstand pH corrosion and maintain most consistent electrochemical catalytic activity (**Figs. A-17 to 19**). This could be attributed to two mechanisms (**Note A-2**): 1) nickel is inert compared with iron and can withstand OH^- group corrosion (**Fig. A-20**); 2) Ni ion has a smaller ionic radius compared to other transitional metal ions (such as Co and Cu ions), which is preferred to withstand ion insertion (**Fig.

A-21). Both mechanisms were further supported with SEM characterizations of the electrodes (**Fig. A-22**) and ICP–MS analysis of the dissolved Fe^{2+} from the electrodes (**Fig. A-23**) before and after electrochemical tests. These results indicated that PB dissolved after tests under different pHs and repetitive CV scans, while NiHCF didn't show any substantial degradation and maintained the highest stability among the transition metal hexacyanoferrates for PB stabilization.

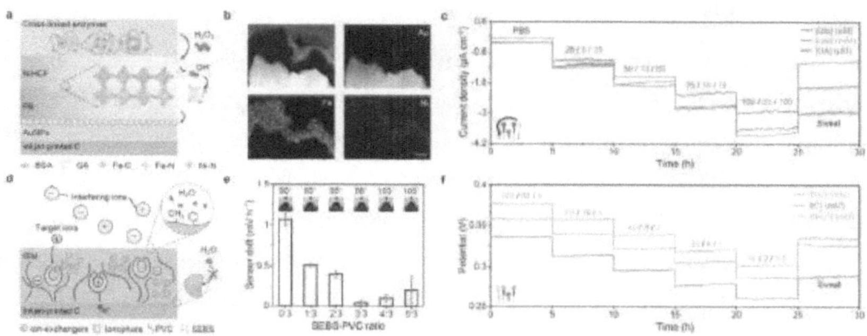

Figure 2-4. Design and characterization of highly robust biochemical sensors. a, Mechanism of enzymatic metabolite sensors. GA, glutaraldehyde. **b**, Cross-sectional STEM and EDS images of the PB-NiHCF interface. Scale bar, 100 nm. **c**, Operational long-term stability of enzymatic glucose, lactate, and UA sensors in PBS and sweat samples for 30 hours. Glu, glucose; Lac, lactate. **d**, Mechanism of ISE sensors. **e**, SEBS-PVC ratios in regards to sensor stability. Insets, contact angle measurements for different SEBS ratios. **f**, Operational long-term stability of ion-selective Na^+, K^+ and NH_4^+ sensors in standard solutions and sweat samples for 30 hours.

Highly stable, continuous, and selective monitoring of sweat glucose, lactate, and UA was realized amperometrically, and a linear response between current output and target concentrations was obtained for all three sensors in physiologically relevant concentration ranges over a 25-hour evaluation period (**Fig. 2-4c**). The sensitivities for glucose, lactate, and UA sensors were 33.65 nA μM^{-1} cm^{-2}, 185.56 nA mM^{-1} cm^{-2}, and 26.36 nA μM^{-1} cm^{-2}, respectively. These sensors also showed a record-breaking long-term stability of more than 100 hours of continuous operation in PBS solutions and untreated human sweat samples, which greatly exceeded those obtained with previous widely adopted wearable sweat sensors (**Figs. A-24 and 25**, and **Table A-1**). It should be noted that as sweat lactate is present in high concentrations (up to 60 mM), an additional diffusion-limited

PVC/DOS membrane was introduced on top of the enzyme film to achieve a wide linear range while maintaining high sensor stability (**Fig. A-26**).

Existing wearable ISEs are based on PVC/DOS membranes and are plagued with a potential drift of typically ~2 mV h^{-1} over time, which is attributed to ionophore leaching and water formation below the ISM[29]. To address this issue during long-term operation, we adopted another analogous composite materials design strategy by introducing SEBS into the PVC system, which shares a similar long-chain structure but holds more methyl and phenyl groups at the sensor interface to promote hydrophobicity and mechanical strength (**Fig. 2-4d**). High hydrophobicity suppresses ionophore leaching and prevents water layer formation at the interface. To fabricate ISEs, the inert yet high surface area nature of inkjet-printed carbon NP electrodes was utilized without the need to deposit additional ion-charge transducer materials. ISM based on the PVC-SEBS matrix were dropcasted onto the carbon electrode, and the ratio of SEBS-PVC was evaluated to identify the optimal stability (**Fig. 2-4e**, **Note A-3**). The optimized ISEs could obtain prolonged stability of 100 hours of continuous operation in both standard solutions and human sweat samples with the potential value decaying less than 0.04 mV h^{-1} (**Figs. A-27 and A-28**, and **Table A-2**). A logarithmic-linear relationship between the potentiometric output of Na$^+$, K$^+$ and NH$_4$$^+$ with near-Nernstian sensitivities of 58.9, 60.6 and 61.2 mV per decade, respectively, was identified during a 25-hour prolonged sensor evaluation in physiologically relevant ranges (**Fig. 2-4f** and **Fig. A-29**).

With an analogous composite materials' approach, our sensors demonstrated high reproducibility (**Fig. A-30**), selectivity (**Fig. A-31**), and long-term continuous operation stability in both standard solutions and untreated human sweat over multiple days (**Figs. A-24, 25, 27 and 28**). Such sensor performance, to the best of our knowledge, was unprecedented in wearable sweat sensing (**Tables A-1 and 2**). The low-cost mass-producible sensor patch is designed to be disposable after use: the anticipated wearable usage time for each patch is 24–48 hours and the users could easily replace the sensor patch. Thus, our sensors can provide a stable response longer than the expected wearable usage time. The general material strategy demonstrated here, based on electrodes prepared by inkjet printing, can be applicable to electrodes manufactured by other scalable technologies including laser engraving and thin-film evaporation (**Fig. A-32**). In addition, the sensor preparation approach here is also not limited to the six sensors we proposed in this study; it can serve as a universal and readily reconfigurable method for other enzymatic and ionophore-based biosensors toward a broad range of practical applications.

2.5 Noninvasive sweat induction and microfluidic sweat collection

2.5.1 Sweat induction through iontophoresis

Iontophoresis gel fabrication: Both anode and cathode of iontophoresis gel were prepared by mixing agarose (3% w/w) into deionized water and then heated to 250 °C under constant stirring until the solution became homogenous. The solution was then cooled down to 165 °C, during which 1% w/w carbachol and 1% w/w NaCl were added to the anode and cathode solution, respectively, and mixed thoroughly. The solution was further cooled down and pooled into the iontophoresis gel reservoirs 41.95 mm^2 for anode and 28.19 mm^2 for cathode, respectively. Together with iontophoresis gels, electrolyte gel (SignaGel, Parker laboratories, INC.) was casted onto the GSR electrodes before placing the CARES device on human subjects.

To realize practical molecular biomarker monitoring without the need for vigorous exercise, miniaturized iontophoresis electrodes coated with carbagels were incorporated into the CARES for autonomous, local sweat induction (**Fig. 2-1a**). Sweat can be continuously secreted from the surrounding glands over a prolonged period of time due to the nicotinic effects of carbachol (transdermally delivered for 5 minutes via a small 50 µA current). Efficient sampling was obtained through custom-developed microfluidics for real-time bioanalysis with high temporal resolution (**Fig. 2-1b, Fig. A-33 and 34**).

2.5.2 Microfluidic evaluation in vitro

On-body flow tests were conducted to evaluate the sweat flow of dual-reservoir designs. An assembled microfluidic patch pre-deposited with black dye in the sweat reservoir was attached to a subject's forearm, followed by *in situ* sweat induction using iontophoresis. Experimental flow tests were also conducted to evaluate the dynamic response of sensors using a syringe pump (78-01001, Thermo Fisher Scientific). Different fluids were injected into the pre-assembled CARES device with a varying flow rate of 1–4 µL min^{-1} (**Fig. A-34**).

2.6 Continuous daily multimodal monitoring

Long-term multimodal sensor evaluation during daily activities: A 24-hour continuous monitoring of physiological and biochemical signals was recorded *via* the CARES device. 5-minute periodical iontophoresis sweat induction was performed at 7:00 am, 9:30 am, 12:30 pm, 4:00 pm, 7:00 pm, 11:00 pm, 1:30 am, and 4:30 am. The physiological data was collected continuously while data from the biosensors were collected 10 minutes after iontophoresis.

Data collection in real life activities: The subject was asked to first perform in-door activities, including relaxation on the phone, playing a long-term VR game (Superhot VR) by wearing a VR headset, and reading journal papers. The subject then performed outdoor activities, including running and walking recovery. The STAI-Y questionnaire was used to assess state anxiety levels during each activity.

Owing to the unprecedented long-term stability of wearable sweat biosensors, the CARES enables long-term real-time continuous monitoring of physicochemical biomarkers. As illustrated in **Fig. 2-5**, the CARES can successfully record the dynamic changes of metabolites and vital signs over 24-hours of activity, involving casual and vigorous exercise, dietary intakes, lab work, relaxing entertainment, and sleep. Glucose and UA levels spiked after food intake, indicating rapidly increased metabolic activities. During vigorous exercise, substantial increases in vascular activity and skin electrolyte/conductivity were observed, and stable output for both metabolites and vital signs were detected during sleep at night. Such powerful capabilities of continuous multimodal monitoring will enable a number of personalized healthcare and human performance monitoring applications.

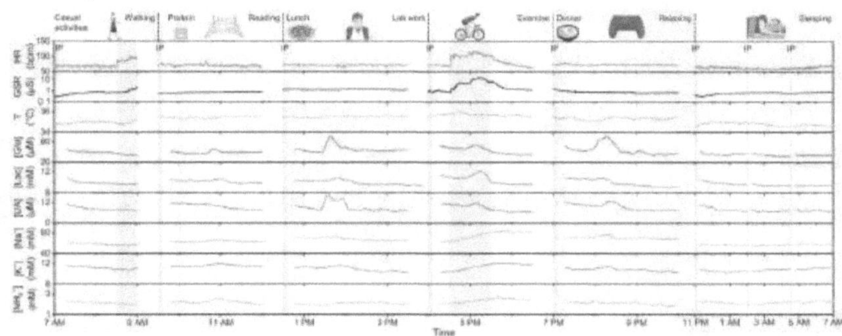

Figure 2-5. Continuous 24-hour multimodal monitoring during a subject's daily activities. IP, iontophoresis; HR, heart rate; bpm, beats per minute.

Chapter 3

BIOFUEL POWERED E-SKIN FOR HUMAN-MACHINE INTERFACES

3.1 Introduction

Recent advances in robotics have enabled soft electronic devices at different scales with excellent biocompatibility and mechanical properties, these advances have rendered novel robotic functionalities suitable for various medical applications, such as diagnosis and drug delivery, soft surgery tools, human-machine interaction, wearable computing, health monitoring, assistive robotics and prosthesis[1-6]. E-skin can possess similar characteristics to human skin such as mechanical durability and stretchability and the ability to measure various sensations such as temperature and pressure[7-11]. Moreover, e-skin can be augmented with capabilities beyond those of the normal human skin by incorporating advanced bioelectronics materials and devices. For example, the compliance of soft e-skin enables wearable sensing of biochemicals from the environment and the human body, showing great potential for wearable personalized health monitoring at molecular levels[12-16].

Sophisticated tasks in practical robotic and biomedical applications including multiplexed sensing, on-demand actuation, and wireless data transmission during prolonged operation set high standards for energy consumption[17]. Batteries have currently been used as the primary power source of the majority of wireless e-skin systems; however, existing ones suffer from inadequate long-term continuous usability, especially when electricity is not readily available[18-20]. Soft e-skins that can directly harvest energy from other accessible sources including human motion, body heat, and solar light, are limited by their low power density. Thus they fail to power signal processing circuitry and wireless data transmission (e.g., Bluetooth or radio frequency communication)[21, 22]. Battery-free e-skins utilizing near field communication for wireless power and data transmission with the user interface have been developed recently, however, these approaches are limited by the required proximity for data processing and readout[23, 24].

Human biofluids, such as human sweat, could serve as an ideal and sustainable bioenergy source for powering future e-skin devices[25, 26]. Moreover, they contain a wealth of chemicals that can reflect an individual's health status[27-29]. Recent advances in wearable sweat analysis have enabled a number of fundamental and clinical applications such as non-invasive metabolic monitoring and disease diagnosis[30-32].

Here, we report a battery-free PPES that harvests energy from human sweat through lactate BFCs, performs continuous multiplexed monitoring of key metabolic biomarkers (e.g., glucose, urea, NH_4^+, pH), and wirelessly transmits the personalized information to a user interface via Bluetooth Low Energy (BLE) (**Fig. 3-1A**). Built on an ultra-soft polymeric substrate, the PPES can comply with the skin's modulus of elasticity and conformally laminate on different body parts for accurate biosensing (**Fig. 3-1B** and **C**). Utilizing cross-dimensional nanomaterial integration, the highly efficient and stable nanoengineered BFC array is able to achieve a record-breaking and sustainable high power density, enabling continuous self-powered health monitoring. Based on this, we report the use of BFCs to fully power an e-skin with both multiplexed sensing and wireless data transmission capabilities. The PPES is successfully validated in vivo in a cycling human trial and its use toward non-invasive metabolic monitoring is further evaluated in human subjects involving dietary and nutrition challenges. Moreover, we demonstrate the use of the PPES as a HMI for assistive robotic control: integrated with strain sensors, the self-powered e-skin could wirelessly transmit signals to a user interface to direct a human operator to control a robotic prosthesis. It is expected that this technology will substantially advance both self-powered e-skin and personalized health care.

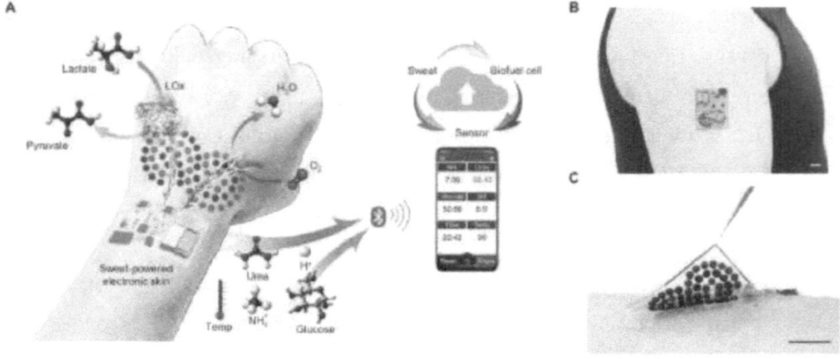

Figure 3-1. PPES for multiplexed wireless sensing. (**A**) Schematic of a battery-free, biofuel-powered e-skin that efficiently harvests energy from the human body, performs multiplexed biosensing, and wirelessly transmits data to a mobile user interface through Bluetooth. (**B** and **C**) Photographs of a PPES on a healthy subject's arm. Scale bars, 1 cm.

3.2 Design of the soft PPES for multiplexed sensing

3.2.1 Materials

Graphite flake, silver nitrate, were purchased from Alfa Aesar. Sodium nitrate, potassium permanganate, hydrogen peroxide, multi-walled CNTs, TTF, citric acid, chloroplatinic acid, sodium borohydride, ammonium chloride, nickel chloride, copper acetate, iron(III) chloride, copper acetate, zinc nitrate, Nafion® perfluorinated resin solution (5%), sodium thiosulfate pentahydrate, sodium bisulfite, PVC, PVB, F127, nonactin, DOS, EDOT, NaPSS, chitosan, PDMS, aniline, platinum carbon black (Pt/C) (0.5%), were purchased from Sigma Aldrich. Sulfuric acid, hydrochloric acid, ascorbic acid, methanol, ethanol, acetone, tetrahydrofuran, sodium chloride, disodium phosphate, urea, dextrose (D-glucose) were purchased from Fisher Scientific. MDB was purchased from Combi-Blocks Inc. LOx (106 U mg^{-1}) was purchase from Toyobo Co. The CNT film and Ni substrate (h-Ni) were purchased from NTL Inc. and MXBaoheng Products, respectively. Medical tapes were purchased from Adhesives Research. Polyimide (PI-2611 and PI-2610) were purchased from HD MicroSystems, Inc. Silver conductive paint was purchased from SPI supplies.

3.2.2 Preparation and characterizations of the BFCs

BFCs, employing the enzymes as biocatalysts to transform the bioenergy into the electricity, are attractive power sources for future wearable and implantable electronics[33, 34]. Among bioenergy resources, lactate, the main metabolic product of both muscle and brain exertion, is present in sweat at tens of millimolar levels and ideally suitable for powering future skin-interfaced electronic devices[35, 36]. However, report of BFCs-powered wearable sensing remain very limited. The main bottlenecks/challenges for the practical use of current BFCs to power e-skin are: 1) Limited power density harvested from the human body (**Table B-1**); 2) Poor BFC stability and short lifetime. Here we utilize a BFC configuration that consists of LOx immobilized bioanodes to catalyze the lactic acid to pyruvate, and Pt alloy NP-decorated cathodes that reduce oxygen to water (**Fig. 3-2A**). Such redox reactions on BFC electrodes yields a stable current to power the electrical loads. Monolithic integration of 0 dimensional (0D) to 3 dimensional (3D) nanomaterials is employed on the BFC electrodes to obtain optimal energy harvesting performance (**Fig. 3-2B** and **C**).

To prepare the bioanodes, h-Ni, rGO films, and MDB-TTF-CNTs are sequentially modified on a Au electrode array (**Fig. 3-2B**). GO suspension was firstly prepared following the modified Hummer's method[48]. Briefly, a mixture of 1 g graphite flake and 23 ml H_2SO_4 was stirred over 24 h, and then 100 mg $NaNO_3$ was added into the mixture. Subsequently, 3 g $KMnO_4$ was added to the mixture below 5 ℃ in the ice bath. After stirring at 40 °C for another 30 min, 46 ml H_2O was added at 80 °C. At last, 140 ml H_2O and 10 ml H_2O_2 (30%, w/v) were introduced into the mixture to complete the reaction. The GO was washed and filtered with 1 M HCl. The self-supported h-Ni was cut into 2-mm-diameter circles using a CO_2 laser cutter and cleaned by ultrasonication in 4 M HCl for 30 min until the color changed from black to silver. After drying, the h-Ni substrates were immersed into a GO suspension with a concentration of 2.0 mg ml^{-1} in water for 1 h. Then the h-Ni substrates were transferred to 5 ml ascorbic acid (10 mg ml^{-1}) overnight and heated at 75 °C for 2 h. After cooling down to room temperature, the rGO/h-Ni composite electrodes were rinsed with water. The free-standing CNTs were immersed into 2 mM MDB solution and then heated to 140 ℃ overnight, followed by rinsing with water for several times. The resulted MDB-CNTs were dropcasted onto the rGO/h-Ni electrode to achieve a higher electrochemically active surface area. The MDB-CNTs/rGO/h-Ni composite was soaked into the 20 mM TTF ethanol/acetone (9:1, v/v) solution. Then bioanodes were obtained by immersing TTF-MDB-CNTs/rGO/h-Ni composite into an LOx solution (20 mg ml^{-1}) for 2 h and dried at 4 °C. 2 µl 0.5% Nafion perfluorinated resin solution was dropcasted on the LOx/TTF-MDB-CNTs/rGO/h-Ni bioanodes to protect the enzymes during the operation.

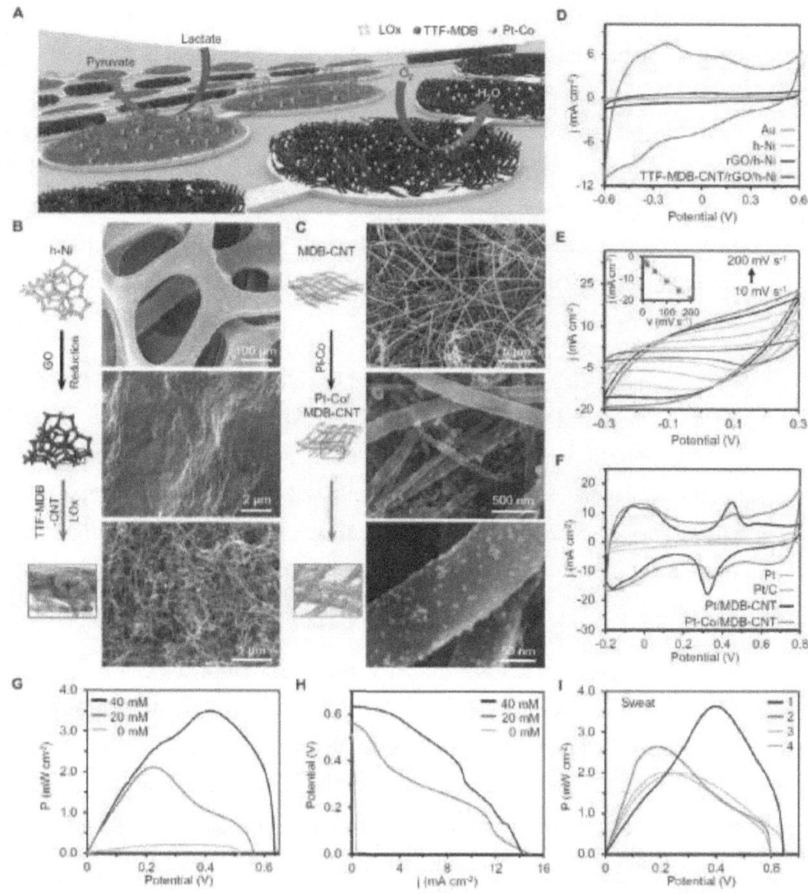

Figure 3-2. Flexible nanoengineered BFC array for efficient energy harvesting. (**A**) Schematic of a soft BFC array consisting of LOx-modified bioanodes and Pt alloy NPs modified cathodes. (**B** and **C**) Schematic and SEM images showing modification process of bioanode assembled with porous h-Ni, rGO, CNTs, and LOx (**B**), and cathode assembled with CNT film and Pt alloy NPs (**C**). (**D**) CV of the Au, h-Ni, rGO/h-Ni, and MDB-

CNT/rGO/h-Ni electrodes. j, current density; scan rate, 50 mV s^{-1}. (E) CV of the MDB-CNT/rGO/h-Ni with scan rates of 10, 20, 50, 100, 150 and 200 mV s^{-1}. Inset, calibration curve of the current density at -0.1 V vs. scan rate (v). (F) CV of the Pt-Co/MDB-CNT, Pt/MDB-CNT, commercial Pt/C, and bulk Pt electrodes recorded in N_2-purged 0.1 M H_2SO_4 solutions. Pt/C, platinum carbon black. Scan rate, 100 mV s^{-1}. (G and H) Power density (P) (G) and polar curves (H) of BFCs recorded in 0 to 40 mM lactate. Experiments in D–H were repeated three times independently with similar results. (I) Power density curves of BFCs in sweat samples from four heathy human subjects. Reference electrodes in D–I, Ag/AgCl.

The extremely high current density and stability of the MDB-TTF-CNT/rGO/h-Ni bioanode illustrated in **Fig. 3-2D** and **Fig. B-1** can be attributed to: 1) the high electrochemically active surface area (ECSA) that is increased by 3000 times after nanomaterial modification (**Fig. 3-2E, Fig. B-2**); 2) the π-π interaction between the CNTs and rGO that enhances the electron transfer rate between LOx and electrodes; 3) the TTF-MDB redox mediator that decreases the overpotential of the lactate oxidation reaction (**Note B-1**).

On the other hand, Pt-based NPs are decorated on an MDB-modified CNT network (MDB-CNT) through electroless plating to form the BFC cathode (**Fig. 3-2C**). The BFC cathodes were prepared as follows: the CNT film was firstly laser cut into 2-mm-diameter disks. The CNT pieces were immersed into 2 mM MDB solution, heated to 140 °C overnight, and then rinsed with water for several times; the MDB-CNT pieces were immersed in 60 mM H_2PtCl_6 solution with 20 mM doping metal ions like Co, Ni, Cu and Zn, and then immersed in a 0.1 M $NaBH_4$ solution for seconds followed by several times of water rinsing; 2 μl 0.5% Nafion perfluorinated resin solution was drop casted onto the Pt or Pt alloy decorated MDB-CNT composite surfaces. It should be noted that the MDB modification is crucial to achieve uniform NP distribution with controlled sizes (**Figs. B-3** and **4**). Compared to that of the conventional bulk Pt electrode, Pt and Pt-Co NPs coated MDB-CNT electrodes show higher ECSA of 170 and 210 times, respectively (**Fig. 3-2F, Fig. B-5**)[37].

The assembled BFC array shows excellent performance, with an open circuit potential (OCP) at ~0.6 V and maximum power outputs of ~2.0 mW cm^{-2} and ~3.5 mW cm^{-2} in 20 and 40 mM lactate solutions, respectively (**Fig. 3-2G and H**). Note that the maximum power densities shift as the lactate concentrations change due to the varied redox reaction of MDB at different pHs (**Fig. B-6**)[38]. When operating in human sweat samples, the BFCs could provide a power density as high as 3.6 mW cm^{-2} (**Fig. 3-2I**). This is, to the best of our knowledge, the highest power density for BFCs in untreated human body fluids (**Table B-1**).

For characterization, the SEM images of the electrodes were obtained by a Field Emission SEM (FEI Sirion). HRTEM images were obtained by a TEM (Tecnai TF-20). Cyclic voltammetry (CV) and LSV analyses were performed through an electrochemical workstation (CHI 660E). The characterizations of BFCs were performed in McIlvaine buffer solutions (pH 6.0) unless otherwise noted. Ag/AgCl references electrodes were used in the BFC characterization. The Pt electrode was fabricated by photolithography (Microchemicals GmbH, AZ 9260) and E-beam evaporation of Cr/Pt (20/100 nm, at a speed of 0.2 Å/s and 0.5 Å/s, respectively), followed by lift-off in acetone. The commercial Pt/C electrode was fabricated by drop-casting 5 μl Pt/C solution (2 mg ml^{-1}) and 2.5 μl Nafion (0.5 %) on the Au electrode (3 mm diameter). To quantify the electrochemically active surface area (ECSA) of the bioanode, the electrodes were tested in McIlvaine buffer (pH 6.0) with different scan rates ranging from 10 to 200 mV s^{-1}. The slope of the calibration curve (the current density at -0.1 V versus scan rate) is used as an estimate of the ECSA of each anode. The adsorption/desorption of hydrogen on Pt (here we use integral area in the CV in the range of -0.2 to 0.1 V) was used to characterize the ECSA of the Pt-based BFC cathode. The polarization curves and power output were achieved by the LSV measurement at a scan rate of 2 mV s^{-1}.

3.2.3 Fabrication and assembly of the soft PPES

The PPES patch consists of two main parts: 1) A nanoengineered flexible electrochemical patch that contains a BFC array and a biosensor array for energy harvesting and molecular analysis in human sweat; biosensing films and biocatalytic nanomaterials are immobilized on serpentine-connected electrode arrays (**Fig. 3-3A**). 2) A flexible electronic patch that consolidates the rigid electronics on an ultrathin polyimide (PI) substrate through flexible interconnects for power management, signal processing, and wireless transmission (**Fig. 3-3B**). A skin-interfaced microfluidic module is integrated into the PPES in order to achieve efficient fresh sweat sampling for stable BFC operation and accurate sweat analysis with high temporal resolution (**Fig. 3-3C**). The independent inlet-outlet design of the microfluidics for BFCs and sensors further minimizes the potential influences of the BFC byproducts on the sensing accuracy. The electronic components and interconnects of the PPES are encapsulated with PDMS to avoid sweat/electronic contact (**Fig. 3-3C**).

Figure 3-3. PPES for multiplexed wireless sensing. (**A** and **B**) Schematic illustrations of the flexible BFC-biosensor patch (**A**) and the soft electronics-skin interface (**B**). (**C**) System-level packaging and encapsulation of the PPES for efficient on-body biofluid sampling. M-tape, medical tape.

The PPES is patterned on an ultra-soft polyimide substrate through standard micro/nanofabrication process to comply with the skin's modulus of elasticity for accurate wearable biosensing. The fabrication process of the PPES platform is illustrated in **Figs. B-11 to 13**. Polyimide (PI-2611) was spin-coated on the silicon handling wafer with a speed of 2000 rpm for 30 s. Then the polyimide was cured at 350 ℃ for 1 hour with a ramping speed of 4 ℃ min^{-1}. The resulted polyimide substrate thickness is about 9 μm. Photolithography (Microchemicals GmbH, AZ 9260) was used to define the inner connection wires. The photoresist was spin-coated on the wafer at a speed of 2400 rpm for 30 s and measured to be around 10 μm thick. For surface treatment, reactive ion etching

(Oxford Plasmalab 100 ICP/RIE, O_2 80 sccm, SF6 5 sccm, 70 W, 20 mTorr) was used for 2 minutes to enhance surface adhesion of polyimide layers. E-beam evaporation of Cu (1.5 μm, at a speed of 2.5 Å s^{-1}) was deposited on the polyimide, followed by lift-off in acetone for minutes. An insulating layer of polyimide (PI-2610) was spin-coated on the surface with a speed of 5000 rpm for 30 s, and then cured at 350 ℃ for 30 minutes with a ramping speed of 4 ℃ min^{-1}. The resulted intermediate polyimide layer thickness is about 1 μm. Another photolithography step was used to define via connections between Cu layers. The wafer was selectively dry-etched using inductively coupled plasma (Oxford Plasmalab 100 ICP/RIE, O_2 50 sccm, 150 W, 80 mTorr, 9 minutes) to form via pattern. Photolithography was used to define outer connection wires. The wafer was surface treated with reactive ion etching using the same recipe described above prior to metal evaporation. E-beam evaporation of Cu (2.5 μm, at a speed of 2.5 Å s^{-1}) was performed and followed by lift-off in acetone. Another encapsulation layer of polyimide (PI-2610) (1 μm thick) was spin-coated and followed by fully curing. Photolithography was performed to define openings of sensors and BFC patterns and then dry etching was performed using inductively coupled plasma (Oxford Plasmalab 100 ICP/RIE, O_2 50 sccm, 150 W, 80 mTorr, 9 minutes).

After wiring system patterning, polyimide was spin-coated on the silicon handling wafer with a thickness of 9 μm. Photolithography (Microchemicals GmbH, AZ 9260) was used to define the shapes of BFCs and sensor arrays. The polyimide was then surface treated with reactive-ion etching to enhance surface adhesion (Oxford Plasmalab 100 ICP/RIE, O_2 80 sccm, SF6 5 sccm, 70 W, 20 mTorr). E-beam evaporation of Cr/Au (20/100 nm, at a speed of 0.2 Å/s and 0.5 Å/s, respectively) was performed, followed by lift-off in acetone. A thin layer of parylene (ParaTech LabTop 3000 Parylene coater) was deposited (1 μm) and followed by photolithography and reactive ion etching (Oxford Plasmalab 100 ICP/RIE, O_2 30 sccm, 100 W, 50 mTorr, 3 minutes) to expose openings for further treatments.

The fully integrated PPES consists of a nanoengineered BFC array (total bioanode area, 1 cm^2), a boost converter, a biosensor array, instrumentation amplifiers, and a programmable system on chip (PSoC) module (integrated with a Bluetooth Low Energy (BLE) module, a microcontroller, and a temperature sensor) (**Fig. 3-4A, Figs. B-9** and **10**). The DC-DC boost converter amplifies the signal potential with a small power loss (~20%) (**Fig. B-9**). The output signal (3.3 V) continuously charges a capacitor (660 µF) that temporarily stores the energy and powers biosensors and other electronic components. The BLE module runs in bursts of activity, periodically waking up from deep sleep mode to acquire measurements with the embedded successive-approximation analog-to-digital converter then wirelessly broadcasting the data to the user interfaces (**Fig. 3-4B**). BLE advertising is chosen here owing to the small size of the data packets and the low power consumption.

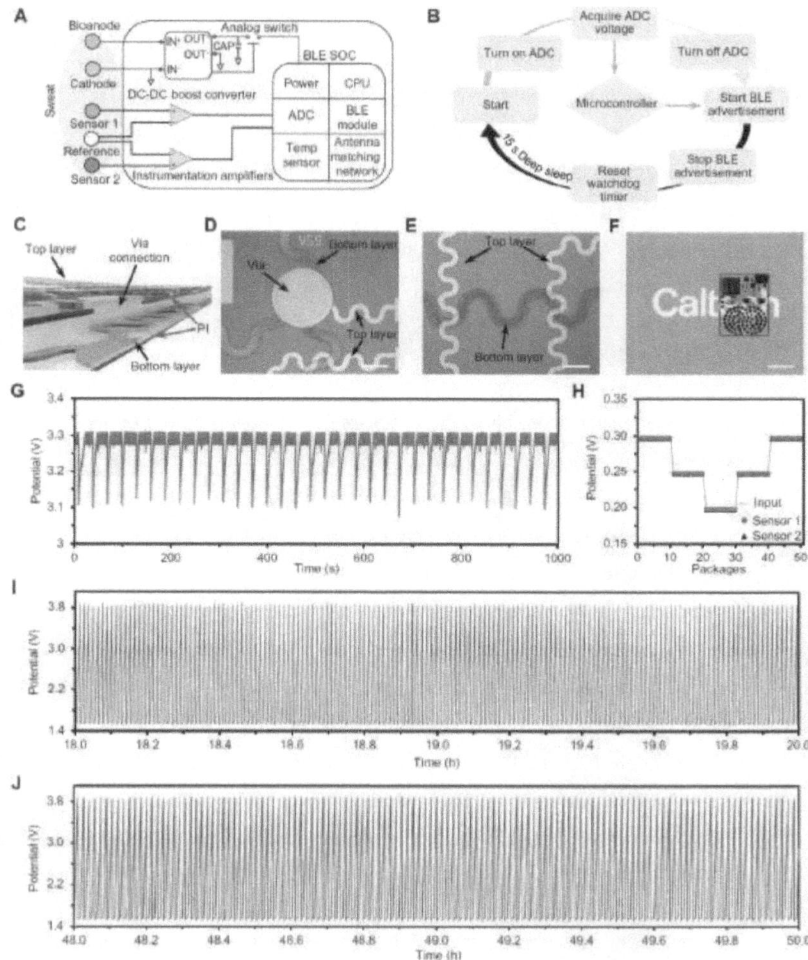

Figure 3-4. System-level integration and evaluation of the PPES. **(A)** Schematic diagram of the PPES system including the BFC array, sensor array, boost converter, instrumentation amplifiers, and Bluetooth Low Energy (BLE) module. SOC, system on a chip; ADC, analog-to-digital converter; CPU, central processing unit. **(B)** Operation flow

of the energy control and data transmission processes. (C–E) Schematic (C) and microscopic images (D, E) showing the via structure and soft interconnects of the e-skin system. Scale bars, 3 μm. (F) A photographic image of a fully integrated PPES. Scale bar, 2 cm. (G) Real-time potential of the capacitor measured during continuous operation of a PPES in 20 mM lactate. (H) Characterization of BFC-powered multiplexed sensing with varied electrical inputs in 20 mM lactate. (I and J) Long-term stability of the capacitor (220 μF) charging process with BFCs through the DC-DC converter over 50 h in 20 mM lactate. Potential dynamics from 18 h to 20 h, and from 48 h to 50 h are illustrated in I and J, respectively. Experiments in G–J were repeated three times independently with similar results.

The packaging and assembly of the PPES are illustrated in **Fig. B-14**. Firstly, the electronic system pattern was set on 100 μm PDMS and connected with the BFC and sensor array pattern by the conductive silver paint (Structure Probe, Inc.). 100 μm PDMS was coated on the BFC and electronic system patterns. The bottom layer was cut as the BFC and sensor array pattern to expose the electrode area. After electrodes were modified on BFC and sensor array, it was combined with the laser-cut microfluidic channels, which were assembled with two medical tape layers and a single PDMS layer in the middle. The fabricated via connection and serpentine-shaped interconnects in the PPES are illustrated in **Fig. 3-4C** to **E**. To minimize the overall patch size, reduce the strain, and achieve uniform strain distribution of the patch during mechanical deformation, the whole BFC electrode array was designed as a serpentine structure (**Fig. B-15**). The fully encapsulated PPES shows excellent transparency, mechanical flexibility and stability (**Fig. 3-4F** and **Fig. B-16**).

The core of the sensor sampling and data processing transmission system is the CYBLE-214009-00 BLE module. This microcontroller provided onboard Bluetooth Low Energy (BLE) capability and 12-bit ADC resolution, as well as a minimal power consumption of 1.3 μA in deep sleep. The open-circuit potential (OCP) of BFCs could hardly reach 1 V, which was significantly lower than the regular electronics needed. Thus, a soft integrated electronic patch, which containing energy boost converter for increasing the applied voltage, multiplexed sensing channels and Bluetooth broadcast was combined with the wearable BFCs. The Texas Instruments BQ25504 Boost Converter forms the core of the circuitry used to harvest energy from the BFCs. The BQ25504 is set to draw current from the BFC at 75% of its OCP to maximize the power provided by the cell. The output of the BQ25504 is connected to a 680 µF capacitor, which is set to charge to 3.3 V before triggering an analog switch to close, connecting the charged capacitor to the rest of the circuitry and powering the system. Then the instrumentation amplifier and microcontroller recorded the response of sensor arrays and sent to the Bluetooth module to broadcast, which were all powered under the boost voltage. The module would run in bursts of activity, periodically waking up in order to take ADC measurements of the sensors, convert the measurement into a voltage, then wirelessly broadcast the data via BLE advertisement before re-entering deep sleep mode. The device was programmed through the Cypress programmable system-on-chip interface using the MiniProg3 SWD debugger and PSoC Creator 4.2 design environment. In order to reduce noise and eliminate the effects of impedance mismatching, instrumentation amplifiers are connected as intermediaries between the sensor electrode output and the microcontroller ADC input. The Texas Instruments INA333 is used as the instrumentation amplifier. The BLE packets broadcasted by the microcontroller are received on a computer with a CY5677 CySmart BLE 4.2 USB Dongle connected. The computer runs a scanner program written in C# with the CySmart 1.3 API which continually searches for an advertising device that matches the BLE address of the CYBLE module, then decodes and stores any received data.

In deep sleep mode, the whole PPES system draws a total current of ~100 µA at 3.3 V, primarily from the two instrumental amplifiers (**Fig. B-17a**). The PPES is programmed to wake up periodically from deep sleep for ~10 ms for acquiring and sending the sensing data with an average consumption of 9.35 mA (**Fig. B-17b**). As a result, the capacitor discharges when the PPES wakes up and will be recharged within a few seconds by the BFCs. In 20 mM lactate, dynamic changes in the potential of the capacitor on the PPES are demonstrated in **Fig. 3-4G**. Wirelessly received BLE data in the user interface shows good agreement with the dual-channel sensor inputs (**Fig. 3-4H**). Our results demonstrate that by using a smaller capacitor (220 or 400 μF), the whole PPES can be continuously powered in lactate solutions (5–20 mM) without the need of deep sleep mode (**Fig. B-18**). Excellent long-term stability of the BFCs-based electrical charging/discharging process is demonstrated by the continuous charging activity for ~60 h, as shown in **Fig. 3-4I** and **J** (and **Fig. B-19**): A capacitor is charged from 1.5 to 3.8 V continuously and repeatedly, and the charging periods could remain stable when fresh lactate fuel is supplied (**Fig. B-19**).

3.2.4 Preparation and characterization of sensors

To prepare the shared reference electrode of the potentiometric sensor array, Ag was first electrodeposited on the Au electrodes with a potentiostati method (-0.25 V for 600 s) in the solution containing 0.25 M $AgNO_3$, 0.75 M $Na_2S_2O_3$ and 0.43 M $NaHSO_3$; the Ag/AgCl electrode was obtained by dropping the 0.1 M $FeCl_3$ solution on top of Ag surface for 60 s; then the PVB reference cocktail was prepared by dissolving 79.1 mg PVB, 50 mg NaCl, 1 mg F127 and 0.2 mg MWCNT in 1 ml methanol; 6.6 μl reference cocktail was modified on the Ag/AgCl electrode and left overnight.

The NH_4^+ ISE was prepared as follows: 100 mg of the NH_4^+ selective membrane cocktail consisting of 1 % NH_4^+ ionophore (nonactin), 33 % PVC and 66 % DOS (w/w) was dissolved in 660 μl tetrahydrofuran. The membrane cocktail was stored at 4 °C. A constant current of 0.2 mA cm^{-2} was applied to electrodeposit the PEDOT:PSS membrane on the Au electrode in the solution containing 0.01 M EDOT and 0.1 M NaPSS to minimize the potential drift of the ISEs. 6.6 μl of the cocktail solution was dropcasted over the PEDOT layer to create an NH_4^+ selective membrane. The modified electrodes were left drying overnight. To prepare urea sensing electrode, 3.7 μL of urease solution (10 mg ml^{-1}) was drop-casted onto the an NH_4^+ ISE for four times, and then 3.3 μL 0.5% Nafion perfluorinated resin solution was dropped over the sensor area. The modified sensors were dried at 4 °C overnight.

To prepare the glucose sensor, differential pulse amperometry (100 cycles in total) was used to electrodeposit the Pt on the Au electrode. -0.4 V potential was applied for 1 s and 1.0 V was used as the cleaning voltage for 0.5 s. 1.1 µl 1% Nafion (prepared by dilution of Nafion perfluorinated resin solution in water) was dropped on the Pt surface, and 2 µl chitosan-GOx mixture (3:1, v/v) was modified on the electrode. The potentiometric glucose sensors were dried at 4 °C overnight; then another 1.1 µl 1% Nafion (1 %) was dropped, covering the enzymes to form the sandwich structure. For the pH sensor, the polyaniline (PANI) was electropolymerized on the Au electrodes in a solution containing 0.1 M aniline and 0.1 M HCl using cyclic voltammetry from -0.2 to 1 V for 50 cycles at a scan rate of 50 mV s^{-1}.

For the in vitro sensor characterizations, analyte solutions were prepared in McIlvaine buffer solutions (pH 6.0 for urea, glucose and NH_4^+). To obtain the best performance for long-term continuous measurements, the NH_4^+ and urea ISEs were placed in a solution containing 0.1 M NH_4Cl and 20 mM urea for 1 h before measurements. The glucose sensors and pH sensors were placed in a solution containing 100 µM glucose and H_2O for 1 h, separately. This conditioning process greatly helps to minimize the potential drift.

CNT-PDMS elastomer was produced by the solution-evaporation method. 7% CNTs (w/w) was added to SYLGARD™ 184 Silicone Elastomer Base and toluene mixture (1:4 v/v) at room temperature. Then the mixture was poured into a culture dish. After toluene evaporated, uncured CNT-PDMS was mixed with curing agent (10:1) and poured onto the mask made by tape and scraped flat with a glass slide. After the mask was removed, the CNT-PDMS was baked at 80 C° for 1 h. Then uncured Ecoflex was spin-coated on it and cured at 80 C° for 1 h. The silver paste was utilized to link the pad on the patterned CNT-PDMS with thin wires. Another layer of Ecoflex was spin-coated for encapsulation and protection. The strain sensors were then connected to the PPES through external wires.

3.3 Results and discussions

3.3.1 Characterization of Pt-Co NP BFC cathode for enhanced stability

Enzymatic BFCs usually suffer from poor long-term stability, primarily because of the limited stability of their cathode. Among commonly used cathode materials, PB, Ag_2O, and MnO_2 are consumable and cannot be recovered in situ[39, 40]. Pt could be easily poisoned during the practical uses due to the formation of the –OH groups on the electrode surface (**Fig. 3-5A**, **Fig. B-7**)[41]; the dissociated oxidized Pt from the NPs tends to redeposit on the larger particles via the Ostwald ripening mechanism (the average cohesive energy of Pt atoms in smaller NPs is smaller than that in larger ones)[42]. To enhance the long-term stability of the Pt particles, transition metal dopants (i.e., Co) are introduced through electroless co-deposition[43, 44]. The resulted Pt/Co alloy NPs are characterized using TEM (**Fig. 3-5B**) and EDS (**Fig. B-8**). The Pt/Co NPs show an average size of ~10–20 nm and an edge lattice spacing of 0.22 nm (**Fig. 3-5B**). The Co dopants could enhance the cohesive energy and thus stabilize the NPs, leading to significantly reduced biofouling in body fluids and higher onset potential for oxygen reduction[45]. In sweat samples, the Pt-Co/CNT shows a relatively stable onset potential compared to that of Pt/CNT (**Fig. 3-5C**). To further improve the long-term stability of the cathode in biofluids, a permselective Nafion layer is modified onto the Pt-Co/CNT (**Fig. 3-5D**). The Nafion/Pt-Co/CNT cathode shows stable performance over 2000 cycles of CV scans (**Fig. 3-5E**) and a negligible fluctuation in the onset potential in a sweat sample for over 30 h (**Fig. 3-5F**). It should be noted that, in addition to high stability, Pt-Co/MDB-CNT cathodes also show highest oxygen reduction performance compared to those modified with Pt or other Pt alloys (**Fig. B-5**), attributed to the Sabatier principle[46].

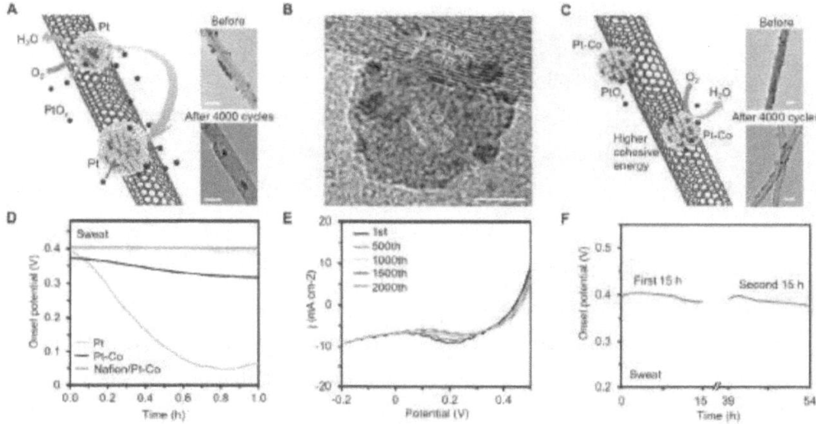

Figure 3-5. Characterization of role of Pt alloy NPs in the BFC stability. (**A**) Schematic illustration showing the Pt NPs (decorated on CNTs) merge into larger ones during the prolonged catalysis. Inset, TEM images of the Pt/MDB-CNT before and after 4000 CV cycles between -0.2 and 0.5 V. Scale bars, 50 nm. (**B**) TEM image showing the crystalline structure of a Pt-Co NP. Scale bar, 5 nm. (**C**) Schematic illustration of the Pt-Co alloy NPs on CNT which maintain the uniform size owing to their higher surface cohesive energy level. Inset, TEM images of the Pt-Co/MDB-CNT before and after 4000 cycles between -0.2 and 0.5 V. Scale bars, 50 nm. (**D**) The onset potentials of Pt/MDB-CNT, Pt-Co/MDB-CNT, and Nafion/Pt-Co/MDB-CNT modified cathodes measured in sweat over a 1-hour period. (**E**) Repetitive linear sweeping voltammograms (LSVs) of a Nafion/Pt-Co/MDB-CNT modified cathode obtained during 2000 CV cycles between -0.2 and 0.5 V. (**F**) Long-term stability test of a Nafion/Pt-Co/MDB-CNT cathode in sweat over 30 h. Reference electrodes in **A**, **C–F**, Ag/AgCl. Experiments in **D–F** were repeated three times independently with similar results.

3.3.2 Characterization of the PPES for multiplexed biosensing

The PPES holds great promise for multiplexed sensing, and a particularly attractive one is wearable sweat analysis for personalized health monitoring. Important sweat biomarkers, such as urea, glucose, pH, NH_4^+, contain meaningful information about an individual's physiological status. Despite high interest, only a limited number of sweat analytes can be accurately monitored by currently reported wearable sensors[12, 13, 47]. Given the complicated sweat secretion process and the electrochemical sensing mechanisms, multiplexed sensing is often essential to achieve an accurate assessment of the specific analyte. As an example, two sweat biosensor arrays (urea and NH_4^+ sensor array, and glucose and pH sensor array) are developed and evaluated toward metabolic monitoring.

The NH_4^+ and urea sensor array is designed on the soft electrochemical patch based on the NH_4^+ ISEs (**Fig. 3-6A**). Compared to the NH_4^+ sensor, the urea sensor contains an additional enzymatic layer where urease converts urea to carbon dioxide and ammonia; the increased NH_4^+ product reflects the urea level. **Fig. 3-6B** and **C** shows the potentiometric responses of the urea and NH_4^+ sensors, measured in 40 to 2.5 mM NH_4^+ solutions and 40 to 2.5 mM urea solutions, respectively. A linear relationship between potential output and logarithmic concentrations of the target analytes is obtained, with near-Nerstian sensitivities of 60.3 mV and 60.0 mV per decade of concentration for NH_4^+ and urea sensors, respectively. The sensors have good selectivity over common analytes in human sweat (**Fig. B-20**). The dependence of urea and NH_4^+ concentrations on the sensor response is illustrated in **Fig 3-6D** and **Fig. B-21**. Considering that NH_4^+ level has a significant influence on urea sensor reading, it is essential to simultaneously monitor both urea and NH_4^+ with real-time calibration for accurate sweat analysis.

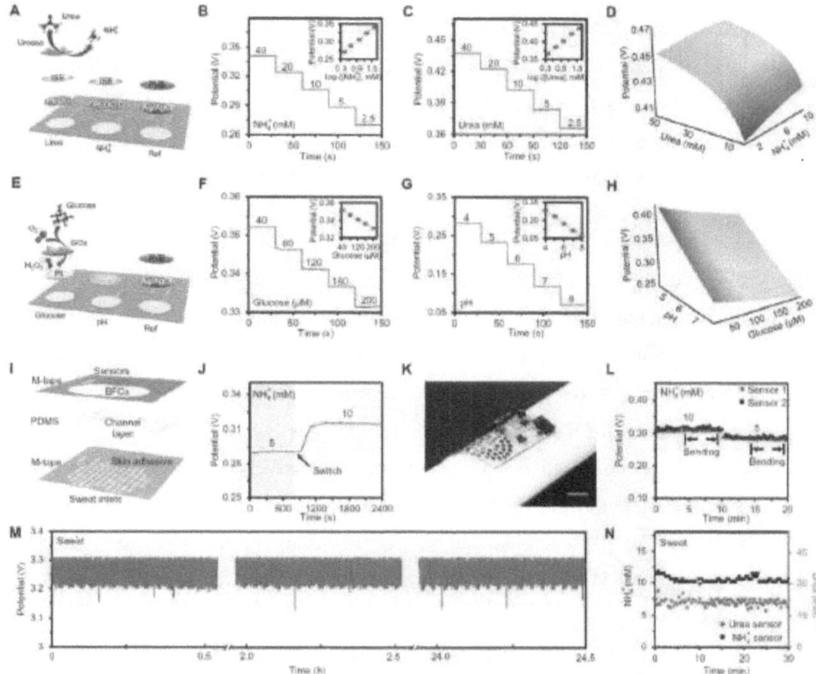

Figure 3-6. Characterization of the PPES for multiplexed biosensing. (A) Schematic of a sensor array for simultaneous urea and NH$_4^+$ monitoring. (B and C) Open circuit potential (OCP) responses of a NH$_4^+$ sensor (B) and a urea sensor (C) in standard analyte solutions. Insets, the corresponding calibration plots of the NH$_4^+$ and urea sensors. Data recording was paused for 30 s for each solution change. Error bars represent the SDs from five sensors. (D) A color map showing the dependence of the urea sensor response on urea and NH$_4^+$ concentrations. (E) Schematic of a sensor array for simultaneous glucose and pH sensing. (F and G) OCP responses of a glucose sensor (F) and a pH sensor (G) in standard analyte solutions. Insets, the corresponding calibration plots of the glucose and pH sensors. Data recording was paused for 30 s for each solution change. Error bars represent the SDs from five sensors. (H) A color map showing the dependence of the glucose sensor on the

glucose and pH levels. (**I**) Schematic of the microfluidic design for biofluid sampling. (**J**) Dynamic response of an NH_4^+ sensor upon switching the inflow solutions at a flow rate of 0.05 ml h^{-1}. (**K**) A photograph of an integrated PPES under mechanical deformation. (**L**) The performance of a two-channel NH_4^+ sensor array under mechanical deformation (with a radius of bending curvature of 1.5 cm). (**M and N**) Potential of the capacitor (**M**) and responses of a urea and NH_4^+ sensor array (**N**) when the PPES operates in a human sweat sample.

The glucose and pH sensor array is prepared using a similar potentiometric sensing approach (**Fig. 3-6E**). A sandwich structure — Nafion/chitosan-GOx/Nafion —is coated on the platinum deposited electrode to form highly sensitive and selective glucose sensor; an electropolymerized polyaniline film serves as the hydrogen ion-selective film for pH sensing. **Fig. 3-6F** and **G** illustrates the responses of the glucose and pH sensors in 40–200 μM glucose and pH 4–8 solutions, respectively. A linear response between the potential output of glucose sensor and glucose concentrations (in physiologically relevant range 0–150 μM) is obtained with a sensitivity of 0.1 mV μM^{-1}. A near-Nerstian sensitivity of 55.3 mV per pH is observed for the pH sensor. Considering that the glucose sensor response is heavily dependent on the solution pH (**Fig. 3-6H**), multiplexed glucose and pH sensing with real-time calibration is also crucial to obtain high sensing accuracy (**Fig. B-22**).

All the sensors show excellent long-term electrochemical and mechanical stabilities during continuous operation, indicating their promise for wearable use (**Figs. B-23 and 24**). Considering that skin temperature has a direct influence on the enzymatic biosensors (glucose and urea sensors here as shown in **Fig. B-25**), the on-chip temperature sensor in the BLE module could provide the skin temperature information for real-time calibration.

For wearable on-body use, the integration of a microfluidic module could greatly enhance the sweat sampling process and lead to a higher temporal resolution for wearable sensing and more stable power output from BFCs (**Fig. 3-6I**). The laser-patterned microfluidics is assembled in a sandwich structure (M-tape/PDMS/M-tape) and contains two reservoirs to minimize the influence of the BFC byproducts on the sensing accuracy (**Fig. B-26**). *In vitro* flow test shows that when the NH_4^+ level in the input solution is switched from 5 to 10 mM at a physiologically measured sweat rate of 0.05 ml h^{-1}, it takes ~4 minutes for the NH_4^+ sensor to reach new stable reading (**Fig. 3-6J**), indicating the small time delay for the on-body continuous monitoring. The PPES is mechanically flexible and can conformally laminate on a curved substrate (**Fig. 3-6K**). Under mechanical deformation with a bending curvature of 1.5 cm in radius, the PPES maintains consistent sensor readings (**Fig. 3-6L**). During long period operation in human sweat samples, the PPES demonstrates stable performance in both energy harvesting (**Fig. 3-6M**) and analyte monitoring (**Fig. 3-6N**).

3.3.3 On-body evaluation of the PPES for wearable sensing

On-body validation of the PPES is conducted on healthy human subjects toward continuous metabolic monitoring during a constant-load stationary biking exercise (**Fig. 3-7A**). **Fig. 3-7B** and **C** shows the data collected from the PPES for real-time monitoring of sweat urea, NH_4^+, glucose, and pH. The pH and NH_4^+ levels measured here, along with the on-chip temperature sensor readings, are also used to calibrate the readings from the glucose and urea sensors, respectively. During the biking process, the urea and NH_4^+ levels in sweat decrease rapidly and then stabilize over time (**Fig. 3-7B**). A similar trend is observed for sweat glucose while a stable pH response throughout the exercise is obtained (**Fig. 3-7C**). The PPES shows good reusability, stability, and biocompatibility during long-term usage (**Fig. B-27**).

In addition to on-body validation, the use of the PPES in metabolic and nutritional management is evaluated through controlled dietary challenges (**Fig. 3-7D** to **I**). **Fig. 3-7D** to **G** demonstrates that, compared to the initial levels, sweat urea and NH_4^+ levels measured 2-hour after a standardized protein intake increase significantly in all three subjects. In contrast, decreased trends are obtained during the 2-hour period from all the subjects without protein intake. In oral glucose tolerance test (OGTT), sweat glucose levels increase dramatically for all subjects after the glucose intake and decrease after 2 hours for subjects without intake. These data indicate the PPES has a great potential in self-powered personalized physiological and metabolic monitoring.

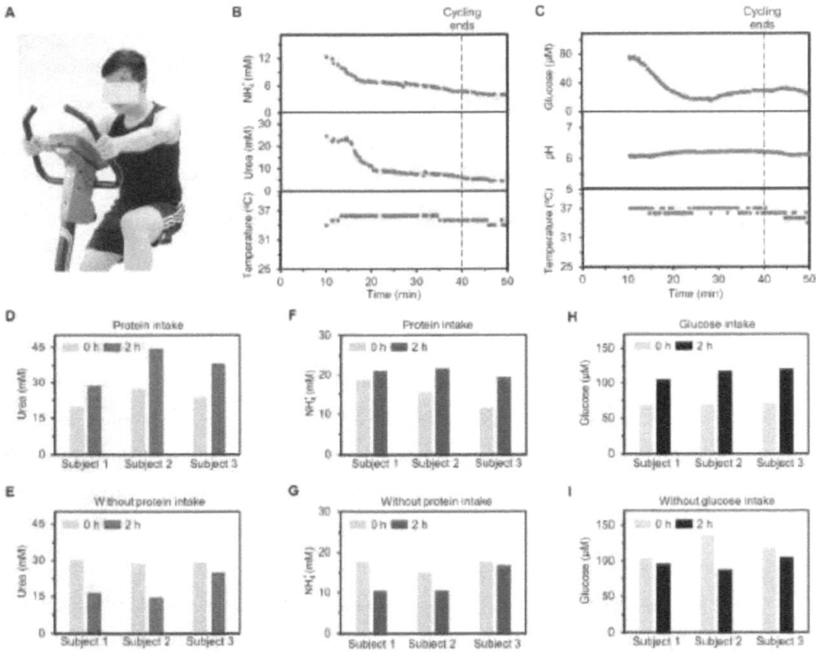

Figure 3-7. On-body evaluation of the PPES toward personalized metabolic monitoring. (A) Photograph of a subject wearing a PPES during cycling exercise. (B) Real-time multiplexed urea and NH_4^+ analysis using a PPES on a subject's forehead. (C) Real-time multiplexed glucose and pH monitoring using a PPES on a subject's forehead. (D to I) Evaluation of the PPES in dietary challenges: sweat urea levels with (D) and without (E) protein intake; sweat NH_4^+ levels with (F) and without (G) protein intake; sweat glucose levels with (H) and without (I) glucose intake.

The validation and evaluation of the lab-on-skin platform were performed using human subjects in compliance with a protocol (ID: 19-0892) that was approved by the institutional review board (IRB) at California Institute of Technology (Caltech). The participating subjects (age range 18–65) were recruited from the Caltech campus and the neighboring communities through advertisement by posted notices, word of mouth, and email distribution. All subjects gave written, informed consent before participation in the study.

In order to validate the PPES, constant-load cycling exercise was conducted on 3 subjects. A stationary exercise bike (Kettler Axos Cycle M-LA) was used for cycling trials. The subjects reported to the lab with overnight fasting. The subjects' foreheads were cleaned with water and alcohol swabs before the PPES was placed. The subjects cycled at 60 rpm for 40 min followed by a 10 min cool down session. When the sweat filled the BFC reservoir, the PPES started to transmit the sensing data to the user interface through the Bluetooth every 15 s, and the data were further converted to the concentration levels after temperature and sweat analyte calibrations. During biking, the sweat would continuously refill both the BFC and sensor array reservoir. For each diary challenge study, after the first cycling trial, the subjects were given a standardized protein drink (30 g protein) for urea and NH_4^+ test, and 3 energy bars (66 g total sugars) for glucose, respectively. After two hours, the subjects cycled at 60 rpm for 40 min followed by a 10 min cool down session. And the PPES would monitor the sweat component change. The first set of data received by the PPES was plotted in **Fig. 3-7D–I**.

3.3.4 Evaluation of the PPES as a HMI for robotic assistance

HMI has attracted substantial attention over the past decade owing to its high promise in real-world biomedical applications in rehabilitation. On skin detection of strains induced by muscle contraction using the soft strain sensors integrated in the e-skin systems is one promising approach to enable dynamic HMI. When integrated with soft strain sensors, the skin-interfaced soft PPES could function as a HMI toward robotic applications (**Fig. 3-8A, Fig. B-28** and **29**). The strain sensors are designed based on CNTs/PDMS elastomer: the resistance of the sensors increase linearly with the applied strain (**Fig. 3-8B and C**). Two strain sensors are placed on the hand and the elbow, respectively, and connected to the PPES (**Fig. 3-8D**); the bending of the finger and elbow could be monitored from the resistance change of the strain sensors (**Fig. 3-8E**). Each resistive type strain sensor as part of a voltage divider consumes a total of ~5 µA. The battery-free e-skin placed on the sweaty arm is able to wirelessly control the motion of a robotic arm in real time: the robotic arm recognizes the gestures of the human arm, then approaches and grabs the target object (**Fig. 3-8F**). The PPES could also be used for robotic assistance in the rehabilitation settings. **Fig. 3-8G and H** demonstrates that prosthesis walking control could be achieved with a strain sensor integrated PPES to achieve assistive walking in real-world environments. By incorporating more physical sensors for EEG and EMG recording along with the continuous metabolic monitoring, the multimodal PPES could facilitate the design and optimization of novel prostheses that bring the human in the loop of prosthesis control to enable real-time user-specific responses to human intent and behavior.

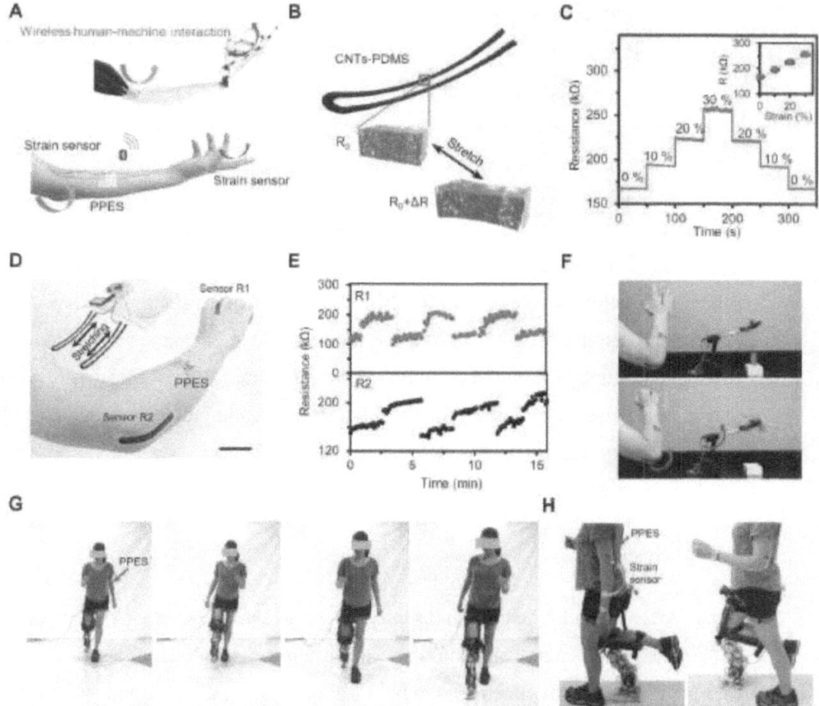

Figure 3-8. On-body evaluation of the PPES as a wireless HMI for robotic assistance. (**A**) Schematic illustration of the PPES for remote human machine interaction. (**B**) Schematic of the CNTs-PDMS elastomer-based stain sensors. (**C**) Resistance response of a CNTs-PDMS strain sensor under different strains. (**D**) Photograph and Schematic (inset) of the PPES integrated with strain sensors. (**E**) Real-time multi-degree motion tracking using a PPES with the strain sensors on subject's finger and elbow. (**F**) Time lapse images of the wireless robotic arm control using a PPES. (**G** and **H**) Time lapse images of front view (**G**) and side view (**H**) of the use of the PPES for robotic prosthesis control.

For robotic arm control, two strain sensors were set on the hand and the elbow of a healthy subject with medical tape, respectively. The subject's arm was cleaned with water and alcohol swabs before the PPES was placed. The subject cycled at 60 rpm for 10 min. When the sweat filled the BFC reservoir, the PPES wirelessly transmitted the strain sensor data to the user interface, and the data were further utilized to control the robotic arm.

For the PPES-based prosthesis control study, a strain sensor was set on the elbow of a healthy subject with medical tape and connected with the PPES. When filled with 20 mM lactate, the PPES wirelessly transmitted the measured strain sensor data to the user interface. When the sensor reached a certain threshold, a human operator sent a computer command through USB to the custom built powered transfemoral prosthesis, AMPRO3[49], to trigger the start of one swing step. With a PD controller, the prosthesis tracked a time-based trajectory designed with the methods in[50]. This research was approved by Caltech IRB with a protocol (ID: 16-0693) for human subject testing.

3.4 Conclusions

Here, we developed a flexible, fully-integrated, and self-powered e-skin platform that can provide real-time, continuous, multiplexed sensing and wirelessly transmit the data to the user interface through Bluetooth communication. Our e-skin platform provides real-time chemical sensing of NH_4^+, urea, glucose, and pH, enhancing the capabilities of e-skin with multimodal and multi-analyte detection. Moreover, the e-skin platform is able to monitor physical parameters such as temperature, strain, and pressure, enabling self-powered wireless human-machine interaction for robotic applications such as powered prosthesis walking. The incorporation of multiplexed, multimodal chemical and physical sensing could expand the potential applications of e-skin for personalized healthcare.

For an e-skin to perform real-time and continuous sensing, onboard powering is necessary. Most e-skin platforms use the battery power supply, which could introduce constraints on the duration of use between charging and affect the long-lasting functionality when the electricity is not readily available. To circumvent the battery requirements, other wireless platforms use battery-free data transmission/powering strategies such as near field communication which require very small readout distance. The use of BLE transmission alleviates such constraints and offers a more realistic communication scheme for e-skin/wearable applications, but would impose a higher power consumption. To accommodate the use of BLE, various battery-free powering strategies have been tested, such as piezo-based or biofuel-based energy harvesting. However, these strategies either have limited power densities for wearable or robotic use which fail to support the BLE functionality on a compact e-skin platform, or suffer from short life-time caused by biofouling.

The PPES proposed here has successfully resolved both challenges for on-board integration of BLE within the compact yet flexible e-skin format. With a unique integration of 0D–3D nanomaterials on bioanodes and cathodes, the BFC designed here yields a record-breaking power density (as high as 3.5 mW cm^{-2} in human sweat) that could support sensing and BLE functionalities on a small e-skin patch. The use of carefully designed and selected Pt-Co alloy NPs has greatly enhanced the long-term stability of the BFC in sweat and enabled the long-term continuous use on skin.

Moreover, the developed e-skin here is a fully-integrated platform with enhanced wearability and sensing accuracy. The platform is ultrathin and transparent, using electronics with low power consumption and integrating electronics onto the soft substrate with a minimal mechanical mismatch, rendering a fully compliant e-skin for maximal comfort and wearability. The use of microfluidics significantly improves sweat refreshing and reduces interference between BFC and biosensors; a carefully studied correlation between different chemical targets yields calibrated sensor readouts that reflect higher accuracy. The system-level integration of electronics, microfluidics and on-board calibration on a soft e-skin provides a much-enhanced accuracy of sensing results for on-body use. Moreover, the successful demonstration of the prosthetic control for robotic assistance using the PPES indicates the great promise of using such platform for human machine interaction and for design/optimization of next-generation prostheses. The development of such a multimodal, fully-integrated and self-powered platform enriches the functionalities and potentials of e-skins and opens the door to numerous robotic and wearable healthcare possibilities such as personalized medicine.

Chapter 4

MULTIMODAL ROBUST E-SKIN FOR WOUND MANAGEMENT

4.1 Introduction

Chronic wounds are characterized by impaired or stagnant healing, prolonged and uncontrolled inflammation, as well as compromised extracellular matrix (ECM) function [1-3]. Over 6.7 million people in the United States alone suffer from chronic non-healing wounds including diabetic ulcers, non-healing surgical wounds, burns, and venous-related ulcerations [4,5], causing a staggering financial burden of over \$25 billion per year on the healthcare system [6]. Chronic wound healing is a highly complex biological process consisting of four integrated and overlapping phases: hemostasis, inflammation, proliferation, and remodeling [1-3]. Current therapies including skin grafts, skin substitutes, negative pressure wound therapy and others can be beneficial, but frequently require procedures or surgical intervention [7]. Microbial infection at the wound site can severely prolong the healing process, and lead to necrosis, sepsis, and even death [3]. Both topical and systemic antibiotics are increasingly prescribed to patients suffering from chronic non-healing wounds, but the overuse, abuse, and misapplication of antibiotics often lead to an escalating drug resistance in bacteria, causing a drastic increase in morbidity and mortality rates [8]. As an alternative therapeutic approach, electrical stimulation has shown to have a significant effect on the wound healing process, including stimulating fibroblast proliferation and differentiation into myofibroblasts and collagen formation, keratinocyte migration, angiogenesis and attracting macrophages [9,10]. However, currently reported electrical stimulation devices usually require bulky equipment and wire connections, making them highly challenging for practical clinical use. More effective, fully controllable, and easy-to-implement therapies are critically needed for personalized treatment of chronic wounds with minimal side effects.

At each stage of healing process, the chemical composition of the wound exudate changes substantially, indicating the stage of healing and even the presence of an infection [11-13]. For example, increased temperature is associated with bacterial infection and changes in temperature can provide information on various factors relevant to healing, inflammation, and oxygenation in the wound bed; acidity (pH) indicates a healing state with balanced protease activities and effective ECM remodeling, moreover, elevated pH in wound environment can be a sign of infection; elevated UA indicates wound severity with excessive reactive oxygen species and inflammation and shows immune system responding to inflammatory cytokines [14]; lactate and ammonium are crucial markers for soft-tissue infection diagnosis and angiogenesis in diabetic foot ulcers [15]; wound exudate glucose has a strong correlation with blood glucose and bacterial activities [16], providing crucial therapeutical guidance for clinical diabetic wound treatment.

Recent advances in digital health and flexible electronics have transformed conventional medicine into remote at-home healthcare [17–23]. Wearable biosensors could allow real-time and continuous monitoring of physical vital signs and physiological biomarkers in various biofluids such as sweat, saliva, and interstitial fluids [18–21,24–30]. In general, an ideal wound dressing should provide a moist wound environment, offer protection from secondary infections, remove wound exudate, and promote tissue regeneration. Despite the promising prospects opened by the wearable technologies [31–37], major challenges exist to realize their full potential toward practical chronic wound management applications: the chronic non-healing wounds pose high requirement on the flexibility, breathability, and biocompatibility of the wearable devices to protect the wound bed from bacterial infiltrations and infection and modulate wound exudate level; the complex wound exudate matrix could substantially affect the biosensor performance and thus there are few reports on prolonged evaluation of biosensors *in vivo* [13,31]; personalized wound management demands both effective wound therapy and close monitoring of crucial wound healing biomarkers in the wound exudate; The absence of miniaturized user-interactive fully integrated closed-loop wearable systems and the evaluation of such systems *in vivo* impede their practical use.

4.2 Design of the fully-integrated stretchable wearable bioelectronic system

4.2.1 Fabrication of the soft wearable patch

Briefly, a 300-nm-thick sacrificial layer of copper was first deposited on the silicon wafer using e-beam evaporation (CHA Mark 40) at a speed of 2.5 Å s^{-1}, followed by standard photolithography (Microchemicals GmbH, AZ 5214) to define the connection wires. Cr/Au/Cr (1/100/20 nm) was deposited on the sacrificial copper through e-beam evaporation of at a speed of 0.2, 0.5, and 0.2 Å/s, respectively, followed by lift-off in acetone. SEBS (200 mg ml^{-1} in toluene) was then spin-coated with a speed of 300 revolutions per minute (rpm) for 30 s. The SEBS film was cured at 70 °C for 1 h to remove toluene and the resulting SEBS film had a thickness of ~300 μm. The copper sacrificial layer was then chemically removed by immersing the silicon wafer in copper etchant (APS-100) for 12 h. The patch was then picked up by a PDMS stamp and rinsed with deionized (DI) water thoroughly. A thin layer of parylene (ParaTech LabTop 3000 Parylene coater) was deposited (200 nm), followed by photolithography and reactive-ion etching (RIE) (Oxford Plasmalab, 100 ICP/RIE, 30 SCCM of O_2, 100 W, 50 mTorr, 90 s) to expose openings for sensors modifications and pin connections. Laser patterning *via* a 50-W CO_2 laser cutter (Universal Laser Systems, power 20%, speed 50%, points per inch 1,000, and vector mode) was used to define the patch shape and outline. After sensor modification, a water-soluble tape (AQUASOL) was used to pick up the wound patch from PDMS backings for further use.

4.2.2 Drug-loaded hydrogel preparation and characterization

Electroactive hydrogel synthesis and 3D printing. 300 mg chondroitin 4-sulfate was dissolved in 1.14 ml of 1 M NaOH under vigorous stirring. Next, 279 μl 1,4-Butanediol diglycidyl ether cross-linker was added and mixed thoroughly for another 30 min. An Anton Paar MCR302 rheometer equipped with a parallel plate to perform rheological characterization. Dynamic viscosity of the samples was measured as a function of shear rate. 3D hydrogel printing was performed based on a custom-designed 3D printer based on a gantry system (A3200, Aerotech) and a benchtop dispenser (Ultimus V, Nordson EFD). 150 μm nozzles were used for the printing. The pump pressure was set to be 2 pounds per square inch (psi) and the nozzle moving speed was set at 5 mm s^{-1}. The printed hydrogel was placed under 60°C for 60 min to form the crosslinked network of the electroactive hydrogels. The crosslinked hydrogels were then left in DI water at 4°C for 48 h (with water replacement every 12 h) to obtain equilibrium swelling.

Drug loading and release studies. The AMP was loaded into the hydrogel by incubating swollen electroactive gel in 1.5 ml AMP solution (2 mg ml^{-1} in Dulbecco's PBS) in a sealed 12 well-plate under 4 °C for 24 h. Passive as well as electro-stimulated release was examined at room temperature in a Dulbecco's PBS solution using the wearable patch. The AMP release was quantified by measuring fluorescence signals using a Synergy HTX Multi-Mode Reader (BioTek Instruments) spectrophotometer at 570 and 583 nm.

Swelling studies. The initial wet weight of each prepared hydrogel was documented. The samples were then immersed in DI water, and the hydrated samples were temporarily taken out of the water and weighed at 1, 4, 8, 24, and 48 h. The swelling ratio was calculated as the weight gain divided by the original weight before hydration.

4.2.3 Wireless system integration of the wearable patch

A 4-layer flexible PCB with a rounded rectangle (36.5 mm x 25.5 mm) geometry was designed using EAGLE CAD. The sensor patch was interfaced directly underneath the flexible PCB through a rectangular cutout (12 mm x 3.8 mm). The power management circuitry consists of a magnetic reed switch (MK24-B-3, Standex-Meder Electronics) and a voltage regulator (ADP162, Analog Devices). The electrical stimulation and drug delivery circuitry consists of a series voltage reference (ISL60002, Renesas Electronics), an operational amplifier square wave generator circuit (MAX40108, Maxim Integrated), and a switch array (TMUX1112, Texas Instruments). The potentiometric, amperometric, and temperature sensor interface circuitry consists of a voltage buffer array (MAX40018, Maxim Integrated), a switch array (TMUX1112, Texas Instruments), a voltage divider, and an electrochemical analog front-end (AD5941, Analog Devices). A programmable system on chip (PSoC) Bluetooth Low Energy (BLE) module (CYBLE-222014, Infineon Technologies) was used for data processing and wireless communication. The fully integrated wearable device was attached to the mice or rats using a 3M double sided tape and fixed with Mastisol liquid adhesive to enable strong adhesion, allowing the animals to move freely over a prolonged period.

4.3 Characterization of multiplexed biomarker analysis and therapeutic capabilities of the wearable patch

4.3.1 Characterization of adhesion of wearable patch

The wearable patch was attached to chicken skin (2×2 cm) using Mastisol liquid adhesive and 3M double sided tape as described previously. A standard T-peel test was then performed according to American Society for Testing and Materials D1876 using a mechanical tester to evaluate patch adhesion to skin. Tegaderm adhesive (3M) was used as control.

4.3.2 In vitro cell studies

Cell lines. Normal Adult human dermal fibroblast cells (HDFs, Lonza) and Normal Human Epidermal Keratinocytes (NHEKs, Lonza) were cultured under 37 °C and 5% CO_2. Cells were passaged at 70% confluency and a passage number of 3–5 was used for all studies.

In vitro cytocompatibility studies. The electroactive hydrogels were washed and transferred to 24-well cell culture inserts (cell culture on permeable supports). The wells were seeded with HDFs and NHEKs (1×10^5 cells per well). The inserts were then placed in cell seeded 24-well plates and cells were treated with appropriate media and incubated under 37 °C and 5% CO_2 for the course of study. A similar study was performed for wearable patches with the cells directly seeded on the patches.

Evaluation of cell proliferation and viability. A commercial calcein AM/ethidium homodimer-1 live/dead kit (Invitrogen) and commercial PrestoBlue assays (Thermo Fisher) were used to evaluate cell viability and cell metabolic activity, respectively. In the live/dead assay, the samples were imaged with an Axio Observer inverted microscope (ZEISS); live cells were stained green with calcein-AM whereas dead cells were stained red with Ethidium Homodimer-1. Using ImageJ software, cell viability was calculated as the percentage ratio of number of live cells to the number of total cells (live + dead).

In vitro wound healing assay. For *in vivo* wound healing assay (circular wound), first a gelatin solution was prepared by dissolving gelatin in DI water (300 mg ml^{-1}) and filtered with a sterile polyethersulfone syringe filter (0.22 μm in pore size). Then 50 μl of the solution was dropped in the center of each well in 12 well-plates. Prior to cell seeding, the plates were kept at room temperature under sterile conditions to keep gelatin in solid condition. Next, HDF cells with a density of 1×10^5 cells per well were seeded in each well and incubated at 37 °C and 5% CO_2. The inherent thermoresponsive properties of gelatin allowed slow dissolving of the gel into the media, creating a uniform-sized wound in the center of cells adhered to the plate. The media was then replaced by fresh media after 4 h and the wound closure was assessed daily for up to 4 days.

4.3.3 In vitro evaluation of wearable patch's antimicrobial properties

Bacterial cells. ·Methicillin-resistant *Staphylococcus aureus* (MRSA, ATCC® BAA-2313™), *Pseudomonas aeruginosa* (*P. aeruginosa*, ATCC® 15442™), and multidrug-resistant *Escherichia coli* (MDR *E. coli*, ATCC® BAA-2452™) and *Staphylococcus epidermidis* (*S. epidermidis*, ATCC® 12228™) were used for antimicrobial tests.

Minimum inhibitory concentration (MIC). The MIC of TCP-25 AMP against different pathogens was evaluated by measuring bacterial optical density. First, bacteria colonies were grown on agar plates containing 15 g l^{-1} agar and 30 g l^{-1} Bacto™ BD Tryptic Soy Broth (TSB) under 5% CO_2 at 37 °C for 24 h. Next, the colonies were transferred and dispersed gently to TSB media, grown overnight in a shaker incubator at 37 °C. A bacteria solution of 10^6 CFU ml^{-1} was prepared for all antimicrobial tests. For MIC test, 200 µl of bacteria solution in TSB was cultured in 96-well plates in the presence of different AMP concentrations (0, 5, 25, 50, 100, 250, 500, and 750 µg ml^{-1}) and incubated at 37 °C for 24 h. Next, the optical density of the solutions was measured, and the relative optical density (as compared to the optical density of the control sample incubated in the absence of AMP) was reported to calculated MIC.

Colony forming unit test. Electroactive hydrogels with and without TCP-25 AMP were placed in 24-well plates and incubated with 1 ml bacteria solution (10^6 CFU ml^{-1}) in TSB media under 37 °C and 5% CO_2 for 18 h. Next, each sample was removed from bacteria solution, washed gently with Dulbecco's PBS (3X) and then placed in microcentrifuge tubes containing 1 ml Dulbecco's PBS. The tubes were vortexed vigorously at 3000 rpm for 15 min to release bacteria trapped inside the hydrogels. A series of logarithmic dilutions (10, 10^2, 10^3, and 10^4) was then prepared from each solution. 20 µl diluted solutions were then seeded on agar plates, followed by incubation under 37 °C and 5% CO_2 for 18 h. The number of colonies was then recorded and reported as colony forming units.

Zone of inhibition. A 100 µl bacteria solution (10^6 CFU ml^{-1}) was dispersed uniformly each agar plate. Next, sterilized electroactive hydrogel disks (6 mm in diameter) loaded with AMP or without AMP were placed into 9 mm holes created in agar plates. The zone of inhibition was measured after 18 h.

4.3.4 Numerical electrical field simulation

Simulation of the electric field generated during electrical stimulation was conducted by using the commercial software COMSOL Multiphysics through finite element method (FEM). Tetrahedral elements allowed modeling of the electric field in 3D space with testified accuracy. The electric field is described by:

$$\nabla \cdot \boldsymbol{D} = \rho$$

$$\boldsymbol{E} = -\nabla V$$

where D, ρ, E and V denote the electric displacement field, charge density, electric field, and electric potential.

The device was fixed at the middle of a cubic computational domain. The side length of the computational domain was 100 mm. The relative permittivity above and below the device was set to be 1 and 76.8, respectively. The boundary condition for the computational domain was set by:

$$n \cdot D = 0$$

where n indicates the normal-to-surface of the boundary. The potential of the anode was set to be 1 V and the potential of the ground electrode was set to be 0 V.

4.4 Evaluation of the wearable patch in vivo for multiplexed wound biomarker monitoring and therapeutic efficacy

4.4.1 In vivo biodegradation and biocompatibility

To assess biodegradation and biocompatibility of the wearable patch, a rat subcutaneous implantation model was used. After anesthesia and analgesia using 2.5% (v/v) isoflurane, 1 mg kg^{-1} Buprenorphine, 5 mg kg^{-1} Ketoprofen, and 1 mg kg^{-1} Bupivacaine, 10 mm incisions in dorsal skin were created to form subcutaneous pockets on the back of Wistar rats (200–250 g, Charles River Laboratories, Wilmington, MA, USA). Next, samples were implanted into each pocket according to the protocol approved by the Institutional Animal Care and Use Committee (IACUC) (Protocol No. IA20-1800) at California Institute of Technology. Animals were then euthanized after 14 and 56 days, and the samples were explanted with their surrounding tissues for further analysis.

4.4.2 Multiplexed wound biomarker monitoring in vivo

The on-body multiplex wound biomarker monitoring was performed using a diabetic wound model in db/db mice (BKS.Cg-Dock7^{m} +/+ Lepr^{db}/J mice, The Jackson Laboratory,

Bar Harbor, ME, USA). After anesthesia and analgesia, a 10 mm full thickness wounds (through to the level of the panniculus carnosus muscle) was created on the dorsum of mice using a surgical blade. A silicon 12 mm-diameter splint (GRACE BIO-LABS, Bend, OR, USA) was placed on the wound area, secured with cyanoacrylate glue and then fixed using Ethilon 5-0 sterile sutures (Nylon). The wearable patch was then placed on the wound and secured on the wound area using Tegaderm™ transparent film dressing (3M). The data from the wearable patch was wirelessly recorded. In the case of the infected wound, a mixture of bacteria solution (50 μl solution, 10^6 CFU ml^{-1} MRSA and 10^6 CFU ml^{-1} *P. aeruginosa*) was applied into the wound area on day 4 post-surgery. For the fasting experiments, the animals were fasted for 24 h (only water was provided). One group of fasting animals were used for injection study. In this case, a 400 mM glucose solution in DBPS (based on body weight) was administration into the mice tail-vein to spark ~10 mM increase in blood glucose level. The *in vivo* sensor readings from the wearable patch were obtained from 30 min prior to injection and continued until 270 min post injection. For the fasting/feeding experiment, the animals were fasted for 24 h, followed by feeding with protein rich laboratory rodent diet 5001 (LabDiet). For the food feeding study, the wearable patch was tested before fasting, after 24-hour fasting and 6 hours fasting/feeding.

4.4.3 Spatial and temporal monitoring of critical-size wounds

Similar to multiplexed wound biomarker monitoring, critical-size wounds (35 mm in diameter) were created in ZDF obese fa/fa diabetic rats (The Jackson Laboratory, Bar Harbor, ME, USA). Next, the sensor array patch was applied on the wound and secured by using 3M Tegaderm™ dressing. Simultaneous sensor readings were recorded daily for both infected and non-infected wounds before and after treatment. For the infection, a similar mixture of bacteria solution (100 μl solution, 10^6 CFU ml^{-1} MRSA and 10^6 CFU ml^{-1} *P. aeruginosa*) was applied into the wound area on day 2 post surgery. During the *in vivo* trial, the data from the wearable patch was wirelessly recorded.

4.4.4 Evaluation of wearable patch-facilitated chronic wound healing in vivo

A 10 mm full thickness wound was created in the ZDF obese fa/fa rat's dorsal skin and the wearable patch was placed on the wound. Four different rat groups were tested with different treatments offered by the wearable patch: negative control, drug release, electrical stimulation, and combination therapy. The animals were euthanized, and the tissue samples were explanted on day 4 and 14 post surgery and processed for further analysis. The adhesion of patches on animals during the course of study was monitored.

4.4.5 In vivo antimicrobial, histological and immunohistofluorescent evaluations

For *in vivo* biocompatibility assessment, upon explantation, samples were fixed in 4% paraformaldehyde under 4 °C overnight, washed thoroughly with Dulbecco's PBS (5X), and then incubated in 30% sucrose overnight (4 °C). The samples were then mounted in optimal cutting temperature compound (OCT compound, Fisher Scientific) followed by flash freezing in liquid nitrogen (N_2), and cryosectioning (10 μm sections). Hematoxylin and eosin (H&E) and immunohistochemistry (IHC) staining were performed on cryosections. For IHC staining, two primary antibodies (anti-CD3 [SP7] (ab16669) and anti-CD68 (ab31630), Abcam) and two secondary antibodies (Donkey anti-mouse, Alexa Fluor 568, and goat anti-Rabbit, Alexa Fluor 488 conjugated antibodies, Invitrogen) were used. Upon antibody staining, the samples were counterstained against DAPI for cell nuclei visualization. The stained slides were then mounted with ProLong™ Diamond Antifade Mountant (Invitrogen) and imaged using a LSM 800 confocal laser scanning microscope (ZEISS).

For regeneration studies, the bacteria samples were first isolated from the wound bed and assessed *via* CFU assay as described earlier. The wound samples were then explanted with the adjacent tissue, processed, sectioned, and stained via Masson's trichrome (MT) staining and IHC. For IHC staining, different primary antibodies including recombinant Anti-Cytokeratin 5 antibody [SP27] (ab64081, Abcam), Anti-NF-kB p65 (phospho S276) antibody (ab194726, Abcam), Human/Mouse/Rat PTEN Alexa Fluor® 647-conjugated

Antibody (IC847R, R&D Systems) and Cytokeratin 14 Monoclonal Antibody (LL002, ThermoFisher) and similar secondary antibodies were used. Upon staining, the samples were mounted with antifade mountant and visualized with a confocal microscope.

4.4.6 Quantitative real-time PCR analysis

RNA was isolated from wound tissue samples using RNeasy Plus Micro Kit (QIAGEN). The RNA quantity and quality were assessed using a NanoDrop 2000/2000c Spectrophotometer at 260/280 nm wavelengths. Next, the complementary DNA (cDNA) was synthesized using QuantiTect Reverse Transcription Kit (QIAGEN). Gene expression was performed using a TaqMan™ Universal PCR Master Mix (ThermoFisher). TaqMan Array Plates for rat wound healing gene expression were used where a library of genes was screened. The cDNAs synthesized in the previous step were then added to each plate and followed by quantitative analysis using a QuantStudio 3 Real-Time PCR system (Applied Biosystems).

4.5 Results and discussions

To address these challenges, here we introduce a fully-integrated wireless wearable bioelectronic system that effectively monitors the physiological conditions of the wound bed *via* multiplexed and multimodal wound biomarker analysis, and performs combination therapy through electro-responsive controlled drug delivery for anti-inflammatory antimicrobial treatment and exogenous electrical stimulation for tissue regeneration (**Fig. 4-1A** and **B**). The wearable patch is mechanically flexible, stretchable, and can conformally adhere to the skin wound throughout the entire wound healing process, preventing any undesired discomfort or skin irritation. Due to the wound's complex pathophysiological environment, compared to previously reported single-analyte sensing, multiplexing analysis of wound exudate biomarkers can provide more comprehensive and personalized information for effective chronic wound management. In this regard, a panel of wound biomarkers including temperature, pH, ammonium, glucose, lactate, and UA were chosen based on their importance in reflecting the infection, metabolic, and inflammatory status of the chronic wounds. Real-time selective monitoring of these biomarkers in complex wound exudate could be realized in situ using custom-engineered electrochemical biosensor arrays (**Fig. 4-1C**). The wearable system's capabilities of multiplexed monitoring, biomarker mapping, and combination therapy were evaluated *in vivo* over prolonged periods of time in rodent models with infected diabetic wounds. The multiplexed biomarker information collected by the wearable patch revealed both spatial and temporal changes in the microenvironment as well as inflammatory status of the infected wound during different healing stages. In addition, the combination of electrically modulated antibiotic delivery with electrical stimulation on the wearable technology enabled substantially accelerated chronic wound closure.

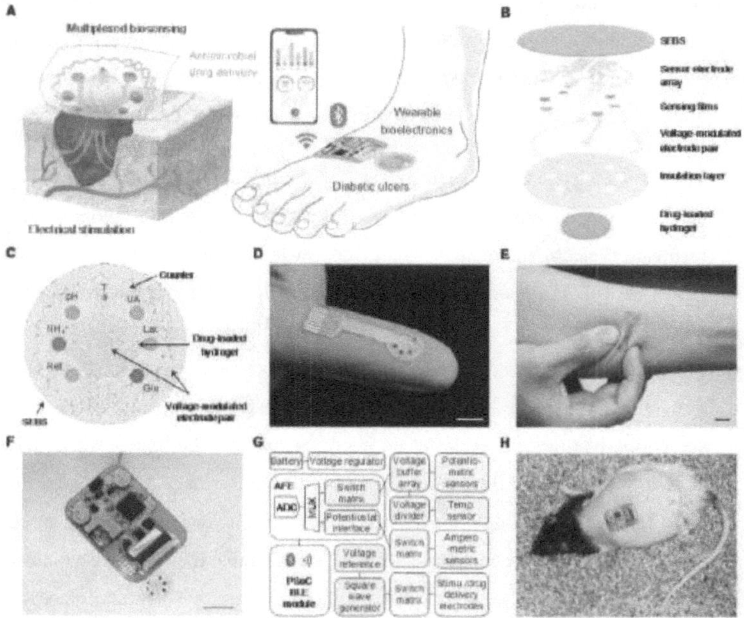

Figure 4-1. A wireless stretchable wearable bioelectronic system for multiplexed monitoring and treatment of chronic wounds. (A) Schematic of a soft wearable patch

on an infected chronic non-healing wound on diabetic foot. (**B**) Schematic of layer assembly of the wearable patch, showing the soft and stretchable poly[styrene-b-(ethylene-co-butylene)-b-styrene] (SEBS) substrate, the custom-engineered electrochemical biosensor array, a pair of voltage-modulated electrodes for controlled drug release and electrical stimulation, and an anti-inflammatory and antimicrobial drug-loaded electroactive hydrogel layer. (**C**) Schematic layout of the smart patch consisting of a temperature (T) sensor, pH, ammonium (NH_4^+), glucose (Glu), lactate (Lac), and UA sensing electrodes, reference (Ref) and counter electrodes, and a pair of voltage-modulated electrodes for controlled drug release and electrical stimulation. (**D** and **E**) Photographs of the fingertip-sized stretchable and flexible wearable patch. Scale bars, 1 cm. (**F** and **G**) Schematic diagram (**F**) and photograph (**G**) of the fully-integrated miniaturized wireless wearable patch. Scale bar in (**F**), 1 cm. ADC, analog to digital converter; AFE, analog front end; PSoC, programmable system on chip; MUX, multiplexer; BLE, Bluetooth Low Energy. (**H**) Photograph of a fully-integrated wearable patch on a diabetic rat with an open wound. Scale bar, 2 cm.

The disposable wearable patch consists of a multimodal biosensor array for simultaneous and multiplexed electrochemical sensing of wound exudate biomarkers, a stimulus-responsive electroactive hydrogel loaded with a dual-function anti-inflammatory and AMP, as well as a pair of voltage-modulated electrodes for controlled drug release and electrical stimulation (**Fig. 4-1B** and C). The multiplexed sensor array patch is fabricated *via* standard microfabrication protocols on a sacrificial layer of copper followed by transfer printing onto a SEBS thermoplastic elastomer substrate (**Figs. C-1** and **2**). The serpentine-like design of electronic interconnects and the highly elastic nature of SEBS enables high stretchability and resilience of the sensor patch against undesirable physical deformations (**Fig. 4-1D** and E). The flexible bandage seamlessly interfaces with a flexible PCB for electrochemical sensor data acquisition, wireless communication, and programmed voltage modulation for controlled drug delivery and electrical stimulation (**Fig. 4-1F–H, Figs. C-3 to 5**). The wireless wearable device can be attached to the wound area with firm adhesion, allowing the animals to move freely over a prolonged period (**Figs. C-6 and 7**).

In addition to the multiplexed and multimodal biosensing, the wearable patch is able to perform combination treatment of chronic wounds through drug release from an electroactive hydrogel layer and electrical stimulation under an exogenic electric field, both controlled by a pair of voltage-modulated electrodes (**Fig. 4-2A–C**). The electroactive hydrogel consists of chondroitin 4-sulfate, a sulfated glycosaminoglycan composed of units of glucosamine, crosslinked with 1,4-butanediol diglycidyl ether (**Fig. C-8**). Due to the shear-thinning behavior of the prepolymer solution, the hydrogel can be precisely fabricated *via* 3D printing (**Fig. C-9**). The negatively charged chondroitin 4-sulfate hydrogel is an ideal choice for loading and controlled release of positively charged large biological drug molecules based on an electrically modulated 'on/off' drug release mechanism (**Fig. 4-2B**). Here, an AMP, TCP-25 [38], was loaded within the chondroitin 4-sulfate hydrogel network through the electrostatic interactions with the polymer backbone, with up to 15% loading efficiency (**Fig. 4-2D**). The highly porous hydrogel network under equilibrium swelling could further enhance the drug loading efficiency (**Fig. C-10**). Under an applied positive voltage, the electroactive hydrogels will be rapidly protonated, resulting in anisotropic and microscopic contraction followed by syneresis/expelling of water from the gel [39], and consequently allowing a controlled release of the TCP-25 AMP (**Fig. 4-2E,F, Figs. C-11** and **12**). In addition, the electrical field will also facilitate the diffusion of positively-charged AMP out of the stimuli-sensitive chondroitin 4-sulfate hydrogel towards the cathode due to electrophoretic flow [40].

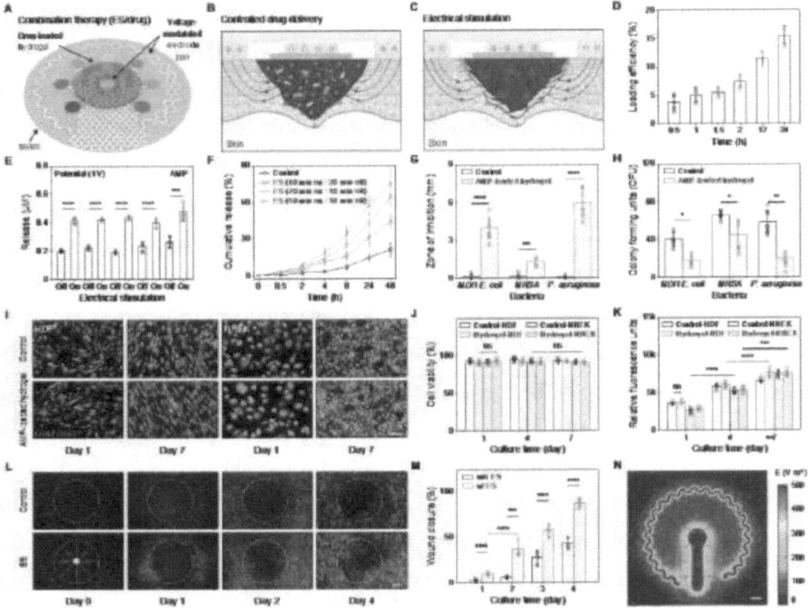

Figure 4-2. Characterization of the therapeutic capabilities of the wearable patch *in vitro*. (A to C) Schematic illustration of the therapeutic modules of the wearable patch (A) and the working mechanisms of the controlled drug delivery for antimicrobial treatment (B) and electrical stimulation for tissue regeneration (C). (D) Loading efficiency of dual-functional TCP-25 anti-inflammatory and AMP into chondroitin 4-sulfate electroactive hydrogel after 0.5–24-hour incubation. (E) Release amount of AMP from the hydrogel under programmed on-off electrical voltage (1 V, 10 min each step). (F) Long term cumulative release of the AMP under programmed electrical modulation. (G and H) *In vitro* antimicrobial tests including zone of inhibition (G) and colony forming units (H) assays for electroactive hydrogels with and without TCP-25 AMP against multidrug-resistant *Escherichia coli* (MDR *E. coli*), *Pseudomonas aeruginosa* (*P. aeruginosa*), and Methicillin-resistant *Staphylococcus aureus* (MRSA). (I to K) *In vitro* cytocompatibility assessment of TCP-25 loaded electroactive hydrogels using live/dead staining (I) and

quantification of cell viability (**J**) and metabolic activity (**K**) for HDF and NHEK cells cultured in the presence of hydrogels. Scale bar, 100 μm. (**L and M**) Fluorescence images (**L**) and quantitative wound closure analysis (**M**) to evaluate wearable patch's therapeutic capability *via* electrical stimulation using an *in vitro* circular wound healing assay created by HDF cells. ES, electrical stimulation. A pulsed voltage was applied for electrical stimulation (1 V at 50 Hz, 0.01 s voltage on for each cycle). Scale bar, 500 μm. (**N**) Numerical simulation of the electrical field generated by the custom-designed electrical stimulation electrodes during operation. E, electrical field. Scale bar, 500 μm. Error bars represent the s.d. (*$p < 0.05$, **$p < 0.01$, ***$p < 0.001$, and ****$p < 0.0001$; $n \geq 3$).

The antimicrobial activity of the TCP-25 AMP-loaded hydrogel was evaluated against gram-positive Methicillin-resistant *Staphylococcus aureus* (MRSA) and *Pseudomonas aeruginosa* (*P. aeruginosa*), and gram-negative multidrug-resistant *Escherichia coli* (MDR *E. coli*) and *Staphylococcus epidermidis* (*S. epidermidis*), the most common pathogenic bacteria associated with microbial colonization of chronic non-healing wounds (**Fig. 4-2G,H** and **Fig. C-13**). The zone of inhibition assay indicates the susceptibility of the MDR *E. coli*, *P. aeruginosa*, and MRSA towards TCP-25 AMP (**Fig. 4-2G**) while the standard colony-forming assay (CFU) showed that the drug-loaded hydrogel was effectively protected from all pathogenic colonization (**Fig. 4-2H**). For cells cultured on AM-loaded hydrogels, the viability of HDF and NHEK cells remained >90% and their metabolic consistently increased during the 7-day culture (**Fig. 4-2I–K** and **Fig. C-14**), indicating that the TCP-25-loaded gels are highly cytocompatible and support cell proliferation.

The wearable patch's therapeutic capability toward enhanced tissue regeneration *via* electrical stimulation was assessed using an *in vitro* wound healing assay (**Fig. 4-2L,M** and **Fig. C-15**). The model wound treated with electrical stimulation showed substantially faster and more consistent migration of HDF cells toward the wound area for 4 consequent days post wounding as compared to the control group without electrical stimulation (**Fig. 4-2L**). Quantitative analysis of the model wound closure indicates higher wound closure rates in the wounds treated with electrical stimulation (**Fig. 4-2M**). The enhanced tissue regeneration is attributed to the directional electrical field generated from our custom-designed electrical stimulation electrodes (**Fig. 4-2N**) which plays a crucial role in cell behavior modulation including cell-cell junctions, cell division orientation, and cell migration trajectories (galvanotaxis or electrotaxis) [41–43]. The electrical potential was applied directly to a pair of insulated electrodes to generate electrical field for electrical stimulation. It should also be noted that continuous electrical stimulation did not cause substantial temperature increase (**Fig. C-16**).

To validate the capability and efficacy of our wearable patch, *in vivo* pre-clinical evaluations are essential. In this regard, the *in vivo* biocompatibility of the wearable patch was assessed. The immunohistofluorescent staining of subcutaneously implanted hydrogel and electrodes in rats showed negligible signs of leukocyte (CD3) and macrophage (CD68) antigens after 56 days, indicating the high biocompatibility of the wearable patch (**Fig. C-17**). All custom-developed biosensors on the wearable patch displayed consistent sensitivity during a 6-h continuous measurement in simulated wound fluid, indicating the high electrochemical stability of the sensors for wound analysis (**Fig. C-18**). *In vivo* multiplexed sensing study was then performed using an infected excisional wound model in diabetic mice. The wound fluid composition was assessed by the wearable patch before infection (day 1), after infection (day 4), and after treatment (day 7) (**Fig. 4-3A**). Substantially elevated UA, temperature, pH, lactate, and ammonium levels were observed as compared to those before infection. The increase in temperature can be potentially linked to inflammation [44]. The elevated levels of UA after infection can be due to upregulation of xanthine oxidase, a component of the innate immune system responding to inflammatory cytokines in chronic ulcers which plays a key role in purine metabolism to produce UA [45]. pH, lactate, and ammonium are all acidity related and their elevation during the bacteria infection has also been widely reported [46]. In contrast, the glucose level in infected wound fluid showed >35% decrease after infection, attributing to the increased glucose consumption of bacteria activities [16]. Upon wound treatment, the temperature, pH, lactate, UA, and ammonium decreased toward the levels prior to the infection while the glucose level increased significantly after treatment, indicating the successful bacterial elimination (**Fig. 4-3A**).

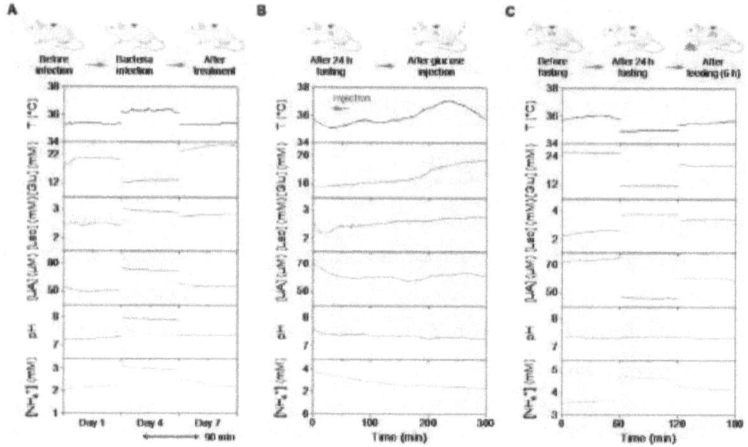

Figure 4-3. *In vivo* **evaluation of the wearable patch for multiplexed wound biomarker monitoring in a wound model in diabetic mice**. (**A**) *In vivo* multiplexed analysis of the chemical composition of wound fluid using a wearable patch in an infected excisional wound model in a diabetic mouse. Infection and treatment were performed after the sensor recording on days 1 and 4, respectively. (**B**) *In vivo* continuous and multiplexed evaluation of wound parameters in a 24-hour fasted mouse before and after glucose administration *via* tail-vein. (**C**) *In vivo* assessment of metabolic changes in wound microenvironment in response to fasting and food feeding in a diabetic mouse.

Considering that dietary intake may have major impact on the composition of diabetic wound fluid, we evaluated the metabolic changes in wound fluid in response to tail-vein glucose administration (**Fig. 4-3B**) and food feeding (**Fig. 4-3C**). Glucose administration *via* tail-vein into the 24-hour fasted mice sparked ~10 mM increase in the blood glucose level. The *in vivo* sensor readings from the wearable patch were recorded from 30 min prior to injection and continued until 270 min post injection (**Fig. 4-3B**). The glucose level in wound fluid showed a gradual increase throughout the 4 hours after injection, indicating a protracted delay with respect to blood glucose. A similar trend was observed for temperature values, attributing to an increased metabolic rate to facilitate digestion. No apparent change in UA level after injection was detected due to the absence of purine intake in the glucose administration. For the food feeding study, the wearable patch was tested before fasting, after 24-hour fasting and 6 hours after feeding (**Fig. 4-3C**). The lactate and ammonium levels increased substantially after fasting while glucose and UA levels decreased after fasting, consistent with the trend of observed blood level changes [47]. In the meantime, temperature decreased due to the fasting-induced hypothermia [48]. As expected, 6 hours after feeding, the glucose and UA levels increased from 11.9 mM to 20.3 mM and from 45.9 µM to 60.3 µM, respectively. These results indicate that wearable patch-enabled wound fluid analysis could be a promising approach to realize continuous and personalized metabolic monitoring.

The wearable patch is mass-producible and readily reconfigurable for various wound care applications. In the case of large chronic ulcers, the wound parameters and microenvironment may vary from site to site, making localized monitoring crucial for optimized assessment and treatment of chronic wound infection. As a proof of concept, we demonstrate customized wearable patches for spatial mapping of physiological conditions of critical-size wounds during the healing process. As illustrated in **Fig. 4-4A** and **B**, we could incorporate a sensor array containing seven pH sensors and nine temperature sensors onto our wearable platform for monitoring and mapping critical-size full-thickness infected chronic wounds in diabetic rats. The pH and temperature sensor arrays showed high reproducibility and stability in simulated wound fluid solutions before and after *in vivo* application (**Fig. 4-4C,D** and **Fig. C-19**).

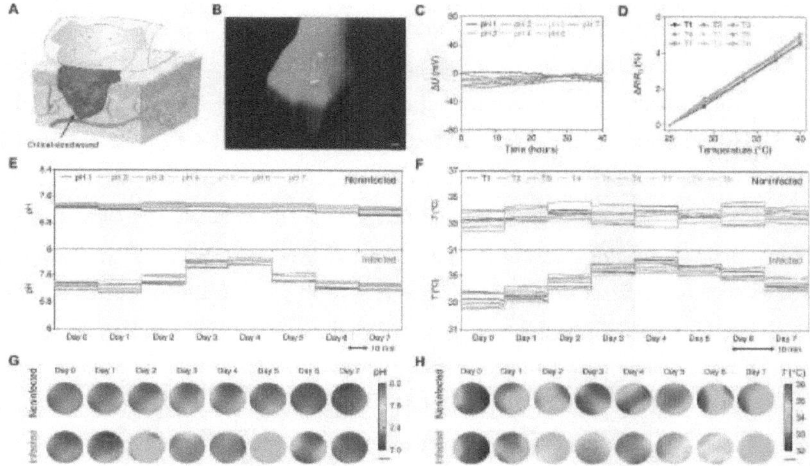

Figure 4-4. Spatial and temporal monitoring of critical-size full-thickness infected wound defects in diabetic rats using the wearable patch. (**A** and **B**) Schematic (**A**) and photograph (**B**) of a soft sensor patch with pH and temperature sensor arrays designed for spatial and temporal monitoring of large and irregular wounds. Scale bar in **B**, 1 cm. (**C** and **D**) The characterization of pH (**C**) and temperature (**D**) sensor arrays on a wearable patch in simulated wound fluid solutions. (**E** and **F**) Dynamic changes in pH (**E**) and temperature (**F**) values of each biosensor on a wearable patch for critical-size non-infected and infected wounds. (**G** and **H**) The mapping of daily local pH (**G**) and temperature (**H**) sensor readings in the wound area for infected and non-infected wounds on each day over the 7-day study period.

On-body validation of the sensor array for spatial and temporal wound monitoring was conducted on critical-size full-thickness wounds (35 mm in diameter) in ZDF rats prior to infection, after infection, and after treatment. The dynamic changes in pH and temperature values for each biosensor on the wearable patch in non-infected and infected critical-size wounds are illustrated in **Fig. 4-4E** and **F**. For non-infected wound studies, the pH and temperature values did not notably change over the 7-day period. However, for infected wound studies, the pH and temperature values increased daily upon applying a mixed infection (MRSA and *P. aeruginosa*) on day 1 and reached the peak value on days 3 and 4. Upon treatment on day 4, the pH and temperature values for each sensor decreased substantially and recovered toward the levels prior to the infection on day 7. The spatial mapping plots of pH (**Fig. 4-4G**) and temperature (**Fig. 4-4H**) in the chronic wound area on each day over the 7-day period were successfully generated based on localized sensor readings. These results are in agreement with previous literature on the changes in the pH and temperature values during the healing progress [46]. A wide variation was observed in both pH and temperature in different regions of the wound upon bacterial infiltration on day 2, showing a higher bacteria growth in the wound edges. The infected wound showed a more uniform pH and temperature at different regions 2- and 3-days post infection due to the formation of uniform biofilm. Upon treatment, the variations increased in the treated wounds on days 5 and 6, indicating the disruption and eventually elimination of the biofilm after treatment (**Fig. 4-4G,H**).

The wearable patch-facilitated combination therapy and wound healing were evaluated in a splinted excisional wound model in ZDF rats (**Fig. 4-5A**). Four different groups were tested: negative control, drug release, electrical stimulation, and combination therapy. The drug treatment was primarily used to eliminate bacterial infections and regulate immune response in early stages of healing. The electrical stimulation was used to facilitate ion channel up-regulation and redistribution, resulting in accelerated cell migration and wound healing. The wearable patch's high flexibility and stretchability provided intact and comfortable contact with the animal's back curvature. Over a 14-day period, the animals were routinely weighed where infected rats showed a non-substantially lower body weight compared to non-infected animals (**Fig. C-20**), indicating that the study procedures did not have any substantial influence on the animals' health. Moreover, the standard CFU on the mixed infection isolated from the wound beds 3 days after drug and combination therapy groups showed a significant reduction in bacterial growth as compared to control and electrical stimulation groups, suggesting effectiveness of the wearable patch in the elimination of pathogenic species from the wound (**Fig. C-21**). Substantially higher rates of wound closure were observed in the treated wounds as compared to the control untreated group, where the group that received combination therapy showed the highest wound closure rate, collagen deposition, and granulation tissue formation, suggesting the recovery of the wound toward the unwounded state (**Fig. 4-5B,C** and **Fig. C22**). We also evaluated the use of the wearable system for multiplexed biosensing and the combination therapy on the same diabetic rats (**Fig. C-23**): compared to the individual evaluation as shown in **Fig. 4-4** and **Fig. 4-5B,C**, similar sensing results and therapeutic effects to the individual evaluation were observed: the continuous sensing data was obtained up to 8 days until the wound dried after therapy while the wound fully closed 14 days post-surgery.

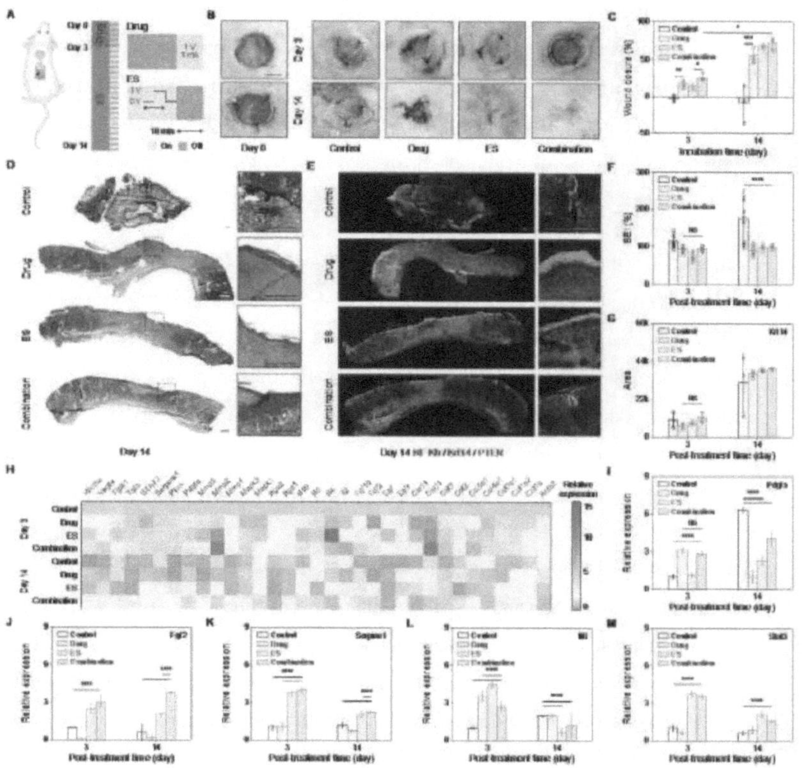

Figure 4-5. *In vivo* **evaluation of wearable patch-facilitated chronic wound healing in full-thickness infected wounds in ZDF rats.** (A) Schematic of the wearable patch on a diabetic wound and the working diagram of combination therapy. (**B** and **C**) Representative images (**B**) and quantitative analysis of wound closure (**C**) for the control wound and wounds treated with drug, ES, and combination therapy on days 3 and 14 post application. Scale bar in **b**, 500 μm. (**D**) Representative images of Masson's trichrome (MTC) stained sections of the full-thickness skin wounds after 14 days of combination treatment. Scale bars, 500 μm. (**E**) Representative immunofluorescent stained images for nuclear factor kappa B (NF-κB) (purple), keratin 14 (Krt14) (green), and phosphatase and tensin homolog (Pten) (red) 14 days after the treatment. Scale bars, 500 μm. (**F** and **G**) Quantitative analysis of scar elevation index (SEI) based on MTC images (**F**) and Krt14 marker based on immunofluorescent images (**G**). (**H**) Quantitative real-time PCR analysis of a library of wound biomarkers for wound biopsies after 3 and 14 days of treatment. (**I** to **M**) Relative expression of Pdgfa (**I**), Fgf (**J**), Serpine1 (**K**), Il6 (**L**), and Stat3 (**M**) genes after 3 and 14 days of treatment. Error bars represent the s.d. (*$p < 0.05$, **$p < 0.01$, ***$p < 0.001$, and ****$p < 0.0001$; n = 3).

In addition, the histopathological analysis of sections of the full-thickness skin wounds *via* Masson's trichrome (MTC) staining (**Fig. 4-5D**) and immunofluorescent staining (**Fig. 4-5E**) were performed. The MTC images showed a significantly higher collagen deposition and granulation tissue formation for treated groups compared to the control group on day 14 (**Fig. 4-5D**). Moreover, the control group on day 14 showed a significantly higher scar elevation index (SEI) of 175±59%, indicating the formation of hypertrophic scars; in contrast, the SEI for combination treatment group was 100±4%, showing uniform dermis repair after treatment (**Fig. 4-5F**). Importantly, the combination therapy was able to accelerate the wound-induced hair follicle neogenesis with adjoining sebaceous glands within the wound bed (**Fig. 4-5D** and **E**, insets) resembling structurally similar glands to those of the uninjured skin [49]. The immunohistochemical analysis of keratin 14 (Krt), a marker of undifferentiated keratinocytes, revealed a delayed re-epithelization in the control group (16%) as compared to the substantially accelerated re-epithelization in combination treatment group (99%) after 14 days (**Fig. 4-5E,G, Fig. C-22**). We further observed a significant growth in the expression of tumor suppressor phosphatase and tensin homolog (Pten), an indicator of higher electrotactic responses [42], among electrical stimulation and combination treatment groups as a direct result of electrical stimulation (**Fig. 4-5E** and **Fig. C-22**). A higher expression of nuclear factor kappa enhancer binding protein (NF-κB), a key signaling factor that promotes remodeling of cellular junctions, cell proliferation, and adhesion [50], was also observed in the combination therapy group on day 14, indicating a higher cutaneous wound healing (**Fig. 4-5E** and **Fig. C-23**).

We further studied the molecular mechanism behind the beneficial effects of our wearable patch's combination treatment on wound healing using quantitative real-time PCR analysis. A library of the most relevant genes associated with wound healing was screened. A substantially elevated expression of growth factors was confirmed in the combination-treated wounds (**Fig. 4-5H, Fig. C-24,** and **Note C-1**) while the control group on day 14 showed a reduced level of these growth factors due to compromised cutaneous wound healing which resulted in impaired re-epithelization and the formation of granulation tissue and ECM. Considering that platelet-derived growth factor subunit A (Pdgfa) plays crucial roles in stimulation of fibroblast proliferation (early function) and induces the myofibroblast phenotype (later function) [51], its elevation supports the higher rate of dermis and granulation tissue formation in the combination treatment group on day 14, while lower or delayed Pdgfa gene expression resulted in impaired wound healing in drug and electrical stimulation groups in the same period (**Fig. 4-5I**). The overexpression of Pdgfa gene in control group after 14 days might be due to the pathogenesis of hypertrophic scars and increased responsiveness of keloid fibroblasts to Pdgf [52]. In addition, the higher expression of fibroblast growth factor (Fgf) genes can be due to higher rate of epidermis regeneration, renewed capillaries, and in cells infiltrating in the granulation tissue [53] (**Fig. 4-5J**). There was also a significant upregulation of serine protease inhibitor clade E member 1 (Serpine1) genes in electrical stimulation and combination treatment groups as compared to the control and drug groups primarily due to the applied electrical field [54] (**Fig. 4-5K**). Serpine1 regulates the extent and location of matrix restructuring and collagen remodeling while facilitating cell motility and proliferation in the process of wound regeneration. Moreover, significant upregulations of proinflammatory cytokine interleukins 6 (Il6) (**Fig. 4-5L**) and signal transducer and activators of transcription-3 (Stat3) (**Fig. 4-5M**) were observed in electrical stimulation and combination therapy groups on day 3. The Il6 can positively influence different processes at the wound site, including stimulation of keratinocyte and fibroblast proliferation, synthesis and breakdown of extracellular matrix proteins, fibroblast chemotaxis, and regulation of the immune response [38], while Stats are cytoplasmic proteins that can transduce signals from a variety of growth factors and regulate target gene expression. Stat3 can be activated upon binding of Il6 to its receptor and thus plays a key role in wound healing [55]. These results further confirmed the powerful combinatorial therapeutic capabilities of the wearable patch to accelerate chronic wound healing.

4.6 Conclusions

We present the development of a wireless wearable bioelectronic system consisting of a multimodal biosensor array for multiplexed monitoring of wound exudate biomarkers, a stimulus-responsive drug-loaded electroactive hydrogel, and a pair of voltage-modulated electrodes for controlled drug release and electrical stimulation. The wearable patch is fully biocompatible, mechanically flexible, stretchable, skin-conformal, and is capable of real-time selective monitoring of a panel of crucial wound biomarkers including temperature, pH, ammonium, glucose, lactate, and UA in multiple rodent models. The wearable patch demonstrated here represents a versatile platform for evaluating wound conditions and intelligent therapy and can be easily reconfigured to monitor several other metabolic and inflammatory biomarkers for various chronic wound care applications.

Despite remarkable progress in developing wearable electrochemical biosensors for continuous monitoring of circulating metabolites in interstitial fluid and human sweat [24,26,56–58], in situ wound fluid analysis remains a major clinical challenge. This is mainly due to the complex and heterogeneous composition of wound fluid (e.g., high protein levels, local and migrated cells, and exogenous factors such as bacteria) leads to severe and unique matrix effects for most previously reported biosensors and failure in accurate measurement of the target metabolite levels in wound fluid [31]. To mitigate such issue, here we introduced the use of an outer porous PU-based membrane that serves as an analyte diffusion limiting layer to protect the electrode, tune response, increase long-term operational stability, linearity, and sensitivity magnitude as well as biocompatibility and mechanical stability of the sensor [59]. Our results indicate that the wearable patch-enabled wound fluid analysis could be a promising approach to realize continuous and personalized wound metabolic monitoring in both a temporal and spatial fashion.

In addition to the multiplexed biosensors, the wearable patch is equipped with an on-demand electro-responsive drug release system, loaded with an antimicrobial and anti-inflammatory peptide. Under an applied positive voltage, the electroactive hydrogels will rapidly release the dual-function peptide which could effectively eliminate bacteria and modulate inflammatory responses in the wound site during the initial stages of healing, in a splinted excisional wound model in diabetic rats. The on-demand drug delivery can be readily modified with different electroactive hydrogels to deliver several other positively or negatively charged drugs and biomolecules (e.g., proteins, peptides, and growth factors). The integration of an electrical stimulation therapeutic module could facilitate cell motility and proliferation, and ECM deposition and remodeling in the process of wound regeneration resulting in rapid and effective cutaneous wound healing.

We demonstrated promising preliminary data for multiplexed *in situ* metabolic monitoring. However, one limitation of the current study is the lack of a continuous wound fluid sampling and circulation system. The mixing of newly secreted analytes with the old ones delayed the sensor response, leading to a compromised temporal sensing resolution. In addition, the long-term continuous operation stability of the biosensors in wound fluid in vivo may need further improvement. Additional anti-fouling protective membranes could be explored to minimize the influence of complex would fluid on sensor performance during *in situ* use. Compared to large critical-size wounds in large animals (e.g., pigs), the spatial biomarker mapping of wounds in rodent models did not reveal substantial spatial variations. Future investigations can focus on employing a microfluidic wound fluid sampling system for efficient capture and continuous delivery of wound fluid to the sensor chamber to improve the temporal resolution of *in situ* biomarker detection [27,56]. Moreover, to improve wearable patch's durability, low-power electronics or energy-harvesting modules could also be implemented into the wearable platform [60–63]. The clinical technology transfer of this product will require multiple further in-depth studies including preclinical biocompatibility evaluation, long-term multiplexed sensor analysis, and efficacy assessment of the closed-loop therapeutic and regenerative modules in pig models due to anatomical, physiological, and functional similarities of pig and human skin wound healing. Further in-depth studies of the cellular and molecular mechanisms behind wound regeneration upon applying the wearable patch's combination therapy *via* single-cell RNA sequencing (scRNA-seq) would be beneficial. Another further device development direction to benefit future evaluation is scale-up manufacturing, packaging, and reliability assessment toward first-in-human studies. We envision that the custom-engineered fully-integrated wearable patch could serve as a more effective, fully controllable, and easy-to-implement platform for personalized monitoring and treatment of chronic wounds with minimal side effects.

Chapter 5

AI-POWERED MULTIMODAL E-SKIN FOR STRESS MONITORING

5.1 Introduction of stress assessment

Stress is a process triggered by demanding physical or psychological events, and may cause anxiety as a prototypical psychological response. While acute stress responses in healthy individuals can be adaptive and manageable, persistent experiences of stress can have deleterious impacts on mental and physical health[1,2], and many mechanisms behind the stress response are yet unknown[3,4] (**Note D-1**). In the United States alone, over 50 million adults suffer from depression, and after the onset of the COVID-19 pandemic, the number of people suffering from mental disorders has drastically risen, causing a heavy burden on the healthcare system[5,6]. Elevated levels of stress and anxiety also pose a large burden to high-demand occupation workers[7], such as athletes[8], soldiers[9], first responders[10], and aviation personnel[11], potentially interfering with their cognitive performance and decision-making process[12]. In response to these impact, understanding and evaluating the stress response has become a cornerstone of clinical healthcare. However, current gold standards for clinical stress response assessments rely on surveys and performance evaluations, which can be highly subjective[13–15]. Thus, there exists a pressing demand for developing a more efficient and effective stress assessment tool that is not characterized by these limitations[16,17].

Recent advances in wearable sensors have enabled real-time and continuous monitoring of physical vital signs[18–22], allowing for a more personalized remote healthcare. Through *in situ* human sweat analysis, wearable biosensors can provide insightful information on an individual's health at the molecular level[23–26]. Despite these promising prospects, major challenges of these sensors exist for clinical applications: a limited set of physical signals are not sufficient for condition-specific assessment of psychological and physiological stress[27]; existing wearable biochemical sensors suffer from poor operational stability in biofluids, which precludes reliable long-term continuous monitoring[28]; the access to human sweat usually requires physical activity that can affect an individual's stress; despite recent progress on stress hormone analysis, continuous monitoring of sweat stress hormones at physiologically-relevant levels using wearable sensor has not yet been achieved due to their extremely low concentrations[29–31]. Therefore, while understanding and monitoring the endocrine response to stress is a promising approach, it is still underdeveloped.

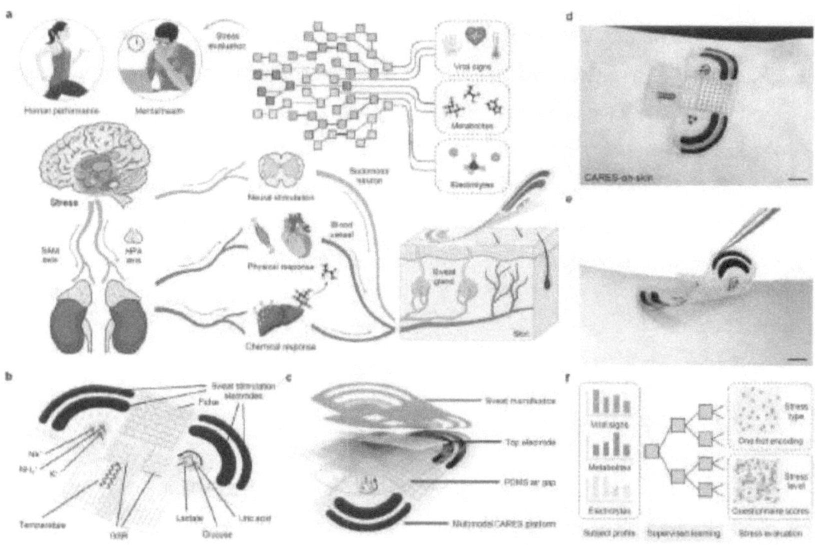

Figure 5-1. CARES platform for stress response monitoring. a, Illustration of the CARES that continuously monitors multimodal physiological and biochemical response from skin, and performs AI-powered stress assessment. **b,** Schematic of the flexible CARES sensor patch and main functionalities: vital sign monitoring, sweat stimulation and sampling, and key metabolite and electrolyte detection. **c,** Schematic of layered structure of the CARES that assembles sensor and microfluidics module. **d,e,** Optical images of a CARES attached to the skin of a human subject. Scale bars, 1 cm. **f,** ML pipeline for CARES-enabled stressor classification and stress/anxiety level assessment.

Non-invasive biomarkers present themselves as a reliable alternative for monitoring the stress response due to the interdependencies between biological and psychological stress. In particular, stress induces a complex biological response within the nervous, endocrine, and immune systems (**Fig. 5-1a**)[32,33]. The perception of stress activates the HPA axis and SAM axis from the hypothalamus in the brain. Acetylcholine in nerve fibers from both axes will stimulate the adrenal gland, releasing stress hormones (e.g., epinephrine, norepinephrine, and cortisol) into the blood. Acetylcholine can also activate sudomotor neurons connected to sweat glands that release ion-rich fluids. This sympathetic activity can be indirectly measured through the GSR and sweat electrolyte levels[34]. The released stress hormones inhibit insulin production, affecting the synthesis of metabolites such as glucose, lactate, and UA, as well as narrow arteries, boosting cardiac activities. By monitoring these stress-relevant biomarkers, it is possible to develop a comprehensive and objective health profile relating biophysical and biochemical signals to dynamic stress response monitoring[35–37].

5.2 System level integration of CARES platform

5.2.1 CARES patch fabrication

Fabricated via a scalable inkjet-printing approach, the wearable device is capable of multiplexed, non-invasive monitoring of key stress-related physiological signals — pulse waveform, GSR, and skin temperature — along with sweat metabolites — glucose, lactate, and UA — as well as electrolytes — Na^+, K^+, and NH_4^+ — during daily activities (**Fig. 5-1b,c**). Through the integration of a miniaturized iontophoresis module, sweat can be induced autonomously at rest without the need for vigorous exercise. Built on an ultrathin flexible polyimide substrate (4 μm) for flexibility and robustness as well as integrated with microfluidics, the CARES device conformally laminates on the wrist for reliable and robust sensing.

The CARES platform consists of a multi-layered sensor patch and a skin-interfaced laser-engraved microfluidic module (**Fig. 5-1d,e**). The sensor patch contains carbachol hydrogel (carbagel)-loaded sweat stimulation electrodes, three enzymatic biosensors, three ISEs, a capacitive pulse sensor, a resistive GSR sensor, and a skin temperature sensor. The platform can be mass-fabricated through serial inkjet printing of silver and carbon as the interconnects and electrodes for top and bottom layers (**Fig. D-1**). A middle PDMS-based airgap layer was spin coated between top and bottom layers, as the soft PDMS facilitates pulse pressure sensitivity and sweat reservoir collection. The microfluidic module was assembled in a sandwiched structure (PDMS/PET/medical tape) and contains two separate reservoirs that enable fresh sweat sampling and rapid refreshing for accurate sweat analysis with high temporal resolution. Carbachol was used for sweat induction as it enables long-lasting sudomotor axon reflex sweat secretion from the surrounding sweat glands owing to its nicotinic effects[38]. In this work, six molecular biomarkers (glucose, lactate, UA, Na^+, K^+, and NH_4^+) were selected as the detection targets due to their strong associations with stress responses (**Note D-2**)[39-43]. Together with laser-patterned microfluidics, the CARES device can be attached to the subject's wrist comfortably and performs multiplexed metabolic sensing *in situ*.

The fabrication process of the CARES is illustrated in **Figs. D-2** and **3**. Polyimide was spin-coated on the silicon oxide wafer at a speed of 5000 rpm for 30 s and then cured at 350 °C for 1 hour with a ramping speed of 4 °C min^{-1}. The resulting polyimide substrate thickness is about 4 μm. For mass-fabrication, 12.5 μm polyimide film was used for large-area patterning demonstration. The CARES patch was then patterned with sequential printing of silver (interconnects and pin connections, reference electrode, pulse sensor, and GSR sensor), carbon (iontophoresis electrodes, counter electrode, temperature sensor, working electrodes for biosensors), and polyimide (encapsulation) using an inkjet printer (DMP-2850, Fujifilm). The CARES patch was then annealed at 250 °C for 1 hour. A 1:12 mixture of curing agent to PDMS elastomer was prepared and stirred thoroughly for 10 minutes, after which the solution was spin-coated at the speed of 800 revolutions per minute (rpm) for 30 s onto the inkjet-printed bottom layer of CARES patch directly, followed by curing at 60 °C for 1 h. The resulting PDMS thickness is about 120 μm. Both the bottom layer and the top layer of the CARES patch were then laser patterned to define outlines and sweat outlets using a 50 W CO_2 laser cutter (Universal Laser System) with power 25%, speed 50%, pulse per inch (PPI) 1000 in vector mode. The bottom layer was further cut to define iontophoresis reservoirs, sweat reservoirs, and airgaps without cutting through the polyimide substrate with an optimized parameter of power 2%, speed 20%, PPI 500 in vector mode for 2 times. The PDMS layer was cleaned with ethanol and deionized water to remove debris, followed by 30 s of O_2 plasma surface treatment using Plasma Etch PE-25 (10 cm^3 min^{-1} O_2, 100 W, 150 mTorr) to clean its surface and promote surface adhesion. The whole CARES patch was then assembled by dry transferring the top layer onto the bottom layer using a PDMS stamp. Biosensors were prepared before microfluidics integration. Note that the sweat reservoir was pre-defined during 120 μm-thick PDMS middle layer in CARES patch fabrication, which has a dimension of 17.15 mm^2 and thereby a reservoir volume of 2.06 μL. The small volume allows a fast refreshing rate and enables rapid detection of dynamic changes during human performance.

5.2.2 Microfluidics module fabrication

The microfluidics layers were fabricated with a laser cutter layer-by-layer by patterning double-sided M-tape, PET, and PDMS with iontophoresis gel reservoirs, gel electrolytes reservoirs, sweat inlets, flowing channels, and outlets. The optimized laser parameters to cut M-tape were set as power 62%, speed 100%, PPI 500 in vector mode for 2 times, and the optimized parameters to cut PDMS were set as power 2%, speed 20%, PPI 500 in vector mode for 2 times to minimize debris. The iontophoresis gel and gel electrolytes reservoirs were patterned by cutting through all microfluidics layers to define gel area and establish gel connection with skin. The first microfluidics layer is a PDMS-based sweat channel layer, which was spin-coated on a PET petri dish and cured at 60 °C for 1 h. The PDMS layer was treated with O_2 plasma before laminating a thin layer of 12 μm PET, followed by laser-defining sweat inlets. Then the third layer of double-sided M-tape was patterned and aligned onto PET, which contacts with the skin and forms the sweat accumulation layer. After attaching the microfluidics module to the CARES patch, the system was further encapsulated with PDMS backings to avoid potential sweat contact and leakage. The device was connected with a flexible printed circuit (FPC) connector for further characterizations.

5.2.3 Microfluidics evaluation in vivo

To realize convenient data collection for real-life applications, in addition to using flexible cables connecting the CARES patch with laboratory instruments (**Fig. D-4**), we further designed a fully integrated wearable CARES system with a flexible PCB for multiplexed and multimodal signal processing as well as Bluetooth wireless communication (**Figs. D-5 to 7**).

5.3 Controlled stressor study designs

5.3.1 Human subject recruitment

The validation and evaluation of the CARES device were performed on healthy human subjects in compliance with the protocols (#19-0892 and #19-0895) that were approved by the Institutional Review Board (IRB) at the California Institute of Technology (Caltech). Participating subjects were recruited from the Caltech campus and the neighboring communities through advertisement by posted notices, word of mouth, and email distribution. 10 healthy subjects (8 males and 2 females, age range 23–38 years) were included in this study. The participants were healthy without anxiety nor depression issues. All subjects gave written informed consent before participation in the study.

5.3.2 CARES on-body protocols

The CARES was mounted on the subject's wrist after skin cleaning with alcohol wipes. Participants were requested to refrain from meals, alcohol, caffeine, and exercise within 3 h prior to the tests. The CARES was sealed in PDMS, leaving output pins exposed with an M-tape backing as support for wire connections. We further designed a plug-and-play input-output to connect with the flexible flat cable (**Fig. D-4**). A 50-μA current was implemented on both pairs of iontophoresis electrodes for 5 minutes simultaneously for sweat induction. The data was collected with a 8-channel multiplexer (CHI Instrument 1430) and a Keithley 4200A-SCS parameter analyzer. A wireless wearable CARES system was also developed for convenient data collection in the real-life settings.

5.3.3 Questionnaire for state anxiety evaluation

State-Trait Anxiety Inventory Form Y (STAI-Y) is a self-evaluation questionnaire that consists of two forms Y-1 and Y-2 to measure state and trait anxiety, respectively, which has a high internal consistency coefficient of 0.91–0.93 for college students and working adults[44]. In our study, we used short form Y-1, which measures state anxiety, as a key psychological response to stress. This measure can be proctored during real-time experiments without major intervention during the stress event. One challenge for quantifying stress is the subjective nature of the questionnaire, which inherently holds a small fluctuations of a couple stress points, with a standard deviation of more than 4 points in most cases [44]. In our study, we take ±2 points as the confidence interval buffer for state anxiety level evaluation.

5.3.4 Stressor protocols in the controlled stress studies

Stressor #1: CPT: The participants were asked to wear the CARES and to relaxation for 10 minutes after iontophoresis sweat induction, during which no sensor signals were collected. After the relaxing stage, both physical and biosensors started monitoring simultaneously as baseline vital and molecular data. The STAI-Y questionnaire was administered to assess state anxiety levels during this relaxed baseline state. The subject was asked to relaxation for another 1000 s, after which a 3-minute CPT was conducted. Subjects were asked to immerse their other hand without the CARES device into a tank containing iced water (0 °C) up to the forearm for 3 minutes. Another STAI-Y questionnaire was then given to evaluate the state anxiety levels and the subjects were asked to finish within 20 s. Afterward, the subjects were instructed to remove the hand from the iced water, and recover in ambient air. Continuous monitoring of multimodal physiological and biochemical data was monitored throughout the stress challenge and recovery stage until 1000 s after the CPT was finished. The subjects were seated during the whole procedure.

Stressor #2: VR test: The sensor data recording process was the same as aforementioned, except that the subjects were asked to play a VR game (Beat Saber) by wearing a VR headset (Oculus Quest 2, Meta). The game was set as one-handed mode with expert difficulty, and the game screen was projected onto a monitor. The subjects were strongly encouraged verbally and asked to compete with other participants' record scores, so that a mixed physical and psychological stress could be stimulated. The STAI-Y questionnaire was used to assess state anxiety levels.

Stressor #3: exercise: For exercise-induced stress, the sensor data recording process was the same as aforementioned. The subject performed the maximum-load cycling (>70 rpm) on a stationary exercise bike (Kettler Axos Cycle M-LA) for 3 minutes or until fatigue, during which strong verbal encouragement was given. The STAI-Y questionnaire was used to assess state anxiety levels.

5.3.5 Results and discussions

To evaluate the use of the CARES for stress response monitoring, controlled experiments were performed on ten healthy subjects using three different stressors, namely the CPT, a VR challenge, and intense exercise (**Note D-3**). The dynamic profiles of all individual sensors integrated in the CARES were collected during each study, as illustrated in **Fig. 5-2** and **Figs. D-8 to 11**. State anxiety levels, as measured by the STAI-Y questionnaire with scores ranging between 10 – 40 points (10 indicating little to no anxiety)[44], were the psychological stress response measure for data training (**Note D-4**). The questionnaire was administered before and after each stressor to quantify the induced anxiety levels within a subject (**Fig. D-12**).

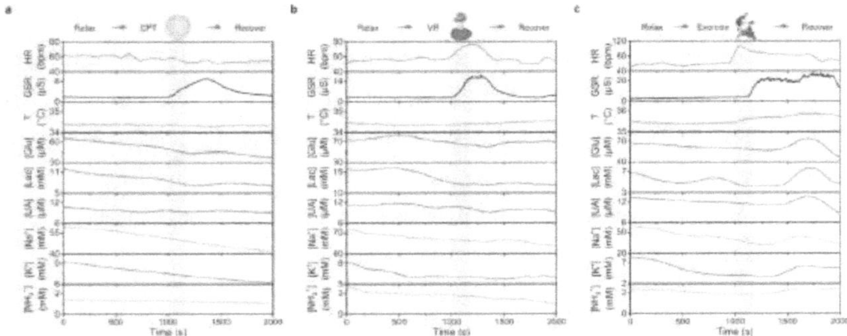

Figure 5-2. On-body evaluation of the CARES under various types of stressors. a–c, Multimodal monitoring of a selected subject's stress response under three different stressors: CPT (**a**) during which the subject was asked to immerse one hand into ice water, VR challenge (**b**) during which the subject was asked to play a VR rhythm game, and cycling exercise test (**c**) during which the subject was asked to perform a maximum-load cycling challenge on a stationary exercise bike.

For each experiment, on-body chemical and physical data showed significant variations in response to each stressor. During the CPT experiment, the subjects immersed one hand in ice water for 3 minutes. A natural reaction of vasoconstriction occurred where the blood vessel constricted in response to cold temperatures[45]. As a result, immediate physiological responses including altered pulse waveform and elevated GSR were observed, consistent with previous reports on the variations of physiological signals with cold-stimulated stress response[46,47]. In addition, delayed mild fluctuations in metabolite concentrations of glucose, lactate, and UA from some subjects were also observed. During the VR test, subjects wore an Oculus VR headset to play a rhythm game (Beat Saber) while the gaming screen was mirrored to a computer monitor with an audience, resulting in both physiological and social-evaluative psychological stress. We observed substantial differences in the pulse waveform and GSR amplitude during and after the stress stimulus, along with elevated glucose, lactate and UA levels minutes later[39–41]. During vigorous exercise, profound activation of the HPA axis led to dramatic changes of all physiological signals as well as sweat metabolites and electrolytes (e.g., Na^+), in agreement with previous studies on exercise induced stress response[42,48]. These results indicate that the CARES can monitor stress-induced biological signals reliably.

5.4 ML-powered stress assessment

5.4.1 Data preprocessing and feature extraction

While all the multimodal sensor signals were monitored in real-time, the data pre-processing was performed asynchronously to extract features. A pulse feature extraction algorithm was developed due to its unique peripheral pulse sampling frequency of $T = 0.007$ s. To match other sensors sampling frequency of $T = 1s$, each pulse waveform was autonomously analyzed through our pulse analysis algorithm, with a floor function afterwards to select the closest pulse feature within each time interval. Signals of the biochemical sensors were manually shifted to align with physical ones due to natural sweat delay. The time stamp when each subject express stress was recorded and manual data labeling was performed. To minimize the variations for inter-subject response, all features were normalized before ML pipeline in regards to each subject during each stress test, in order to generalize the model among population. After data collection and analysis, the training and testing datasets were shuffled and divided 8:2, respectively, and were randomly selected using an equal representation of each class. ML model was developed to link the biological and chemical features to the stress detection, stress types, as well as state anxiety levels from questionnaire scores.

5.4.2 Model selection for stress classification

All training model were built using Python (Python 3.8) based on the data collected from ten subjects facing three different stressors with a set of 60,000 s of CARES recordings. Segmentation of the sensor signals was done using a sliding window with a sampling interval of 1 s, given each stress type representation. A number of ML models were evaluated according to precision-recall curve and their F_1 score, including linear and radial basis function (RBF) support vector machines (SVM), logistic and ridge regression, conventional decision trees, as well as gradient-boosted decision tree XGBoost model. The trained XGBoost model outperformed typical ML models for both stress detection and stress type classification.

5.4.3 Model selection for stress regression

The ML algorithms were developed on a password-protected local computer with individual GPU module Nvidia 3080. The training model were built as aforementioned, except that the kernel was changed to regressor instead of classifier. For overall stress level evaluations, on the other hand, features were extracted from stress region by taking average signal changes from the moving average (MA) of sensors rather than segmented at each timepoint, and simpler ML models such as linear regression and SVM were evaluated due to the reduced size of dataset to prevent overfitting. A brute force examination of features was performed to compare the contributions of physicochemical biomarkers.

5.4.4 Results and discussions

To quantify the stress response-related features, data-driven stress and anxiety evaluation were performed after each experiment was complete, where an ML pipeline was developed to extract features and deconvolute connections between physicochemical information and stressor types and state anxiety levels (**Fig. 5-3a** and **Note D-5**). We undertook this challenge using three separate ML analyses: stress detection versus relaxation, stressor classification, and anxiety level evaluation, where we trained and tested each model across three experiments (VR, CPT, exercise) of all ten subjects for a total of 60,000 s of physiological CARES signals. All signals were calibrated and normalized to ensure that the features extracted after data pre-processing were stable against patch variations and any moderate motion artifacts (**Figs. D-13** to **15**, **Note D-6**, and **Table D-1**). Feature extraction was validated before ML analysis through projecting the multidimensional feature-space into 2D space via t-distributed stochastic neighbor embedding (t-SNE)[49], where data from stress/relaxation naturally formed distinctive clusters, indicating the discriminative power of the features (**Fig. 5-3b**, **Fig. D-16a**).

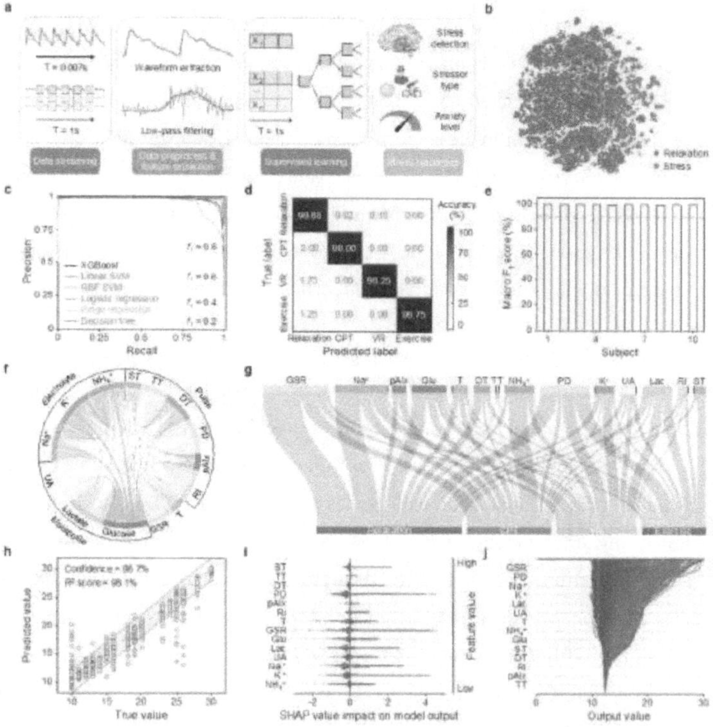

Figure 5-3. ML-powered stress response assessment. a, Schematics of the ML architecture for data preprocessing, feature extraction, supervised learning and evaluation. **b**, t-distributed stochastic neighbor embedding (t-SNE) plot from the dataset recorded by the CARES visually showing feature separation in a 2-dimensional space. **c**, Precision-recall curve of different ML models for stressor classification. XGBoost, extreme gradient boosting; SVM, support vector machine; RBF, radial basis function. **d**, Confusion matrix displaying the classification accuracy for predicting each type of stressor in test set. **e**, The

overall stress classification accuracy based on macro-averaged F_1 score for each subject. **f**, A chord diagram showing the relative correlation between different sensors. ST, systolicTime; TT, tidalPeakTime; DT, dicroticPeakTime; PD, pulseDuration; pAIx, peripheral augmentation index; RI, reflectionIndex. **g**, Sankey diagram of SHAP analysis depicting the relative contribution of different sensors to stressor classification. **h**, True versus the ML-predicted state anxiety scores. ±2 state anxiety score buffer is shown based on the potential error in the anxiety questionnaires. **i**, Shapley additive explanation (SHAP) summary plot for state anxiety level evaluation based on the dataset collected by the CARES. Each axis plots the distribution of SHAP values of a given feature for each prediction instance. **j**, SHAP decision plot explaining how the ML model determines the state anxiety level using both physiological and biochemical features.

Different ML models were evaluated, and the trained boosting decision tree model Extreme Gradient Boosting (XGBoost) outperformed typical ML models, including linear and radial basis function (RBF) support vector machines (SVM), logistic and ridge regression, and conventional decision trees (**Fig. 5-3c**). Combined with features extracted from both physiological and metabolic data, it was found that our XGBoost ML model could yield a much higher accuracy, with stress response classification accuracy of 99.2% for stress/relaxation detection (**Fig. D-16**) and an accuracy of over 98.0% for stressor classification, which to the best of our knowledge is the highest accuracy reported for stressor classification (**Fig. 5-3d**, **Table D-2**). It should be noted that differentiating stressors has high significance, as each stressor carries varying physiological and psychological influences and could act as risk factors for coping responses and cardiovascular diseases[50–52]. Distinguishing types of stressors has been recognized as a necessary condition for understanding the complex interrelationships among distinct stress experiences, as well as the collective impacts of stress on mental health[53] (**Note D-3**). Moreover, it resulted in highly consistent overall accuracies of over 99.3% across different individuals (**Fig. 5-3e**, **Note D-7**).

The Pearson correlation coefficients between all sensors in the CARES show the interrelatedness between physiological and chemical biomarkers (**Fig. 5-3f**). The relatively homogeneous correlation shows the high independence of the extracted features. To evaluate each physicochemical sensor's contribution to the model, feature importance of each biomarker towards each stressor was evaluated using a Shapley additive explanation (SHAP) (**Fig. 5-3g**, **Fig. D-17**, and **Note D-7**). Through SHAP analysis, the feature importance of GSR, pulse, glucose, and Na^+ indicate these biomarkers play a significant role in stressor classification. These results support the fact that stress responses involve participants' vascular dynamics, neural stimulation, and metabolism.

Based on the classification results, we expanded our analysis to state anxiety level evaluation. We adopted a similar XGBoost regression model and could predict state anxiety levels with a high confidence level of 98.7% and 98.1% coefficient of determination of scores from the STAI-Y (with a standard deviation of 4 points or less[44]) (**Fig. 5-3h, Note D-4**). The relevance of each feature was evaluated using SHAP analysis as well (**Fig. 5-3i,j**). Through SHAP analysis, it was determined that GSR, pulse, Na^+, K^+, and lactate played the most important role in state anxiety level prediction. Note that SHAP values show the relative significance of each feature in the ML model.

The wireless system was successfully used for on-body tests and validation of our CARES systems in the laboratory settings (**Fig. D-18**) and in real-life daily casual activities (**Fig. D-19**). Our ML models obtained from the laboratory tests were able to accurately classify the types of stressors and state anxiety levels based on the wirelessly collected sensor data in the laboratory (**Fig. D-20**) as well as real-life settings (**Fig. D-21**). We anticipate that for large-scale human trials, the CARES will surpass the current gold standards for stress response quantification, and provide a highly robust stress response monitoring tool that is not reliant on subjective reporting with its potential for errors. In this regard, we envision a high potential for wearable multimodal physicochemical monitoring of dynamic stress response.

Additionally, given the intrinsic limitations of questionnaires being able to only characterize state anxiety levels within a given time period rather than dynamic stress change continuously, we analyzed the stress response event as a whole to mimic questionnaire functionalities (**Note A-1**). In this circumstance, features were extracted from the stress region by taking mean signal changes from the moving average (MA) of sensor data rather than segmented at each timepoint, and a simple linear regression model was trained with fewer features selected to correspond to questionnaire scores and prevent overfitting (**Fig. D-22**). With the reduced size of dataset and analyzing the overall sensor responses in CARES, we performed a brute force feature selection within each biomarker and found that combined physicochemical features outperformed that of physical and chemical sensors alone.

Chapter 6

CONCLUSION AND FUTURE OUTLOOK

This book has summarized our efforts to introduce multimodal physiological and biochemical sensors into the e-skin. We developed a general approach through materials engineering by applying analogous composite materials for stabilizing and conserving sensor interfaces, and achieved highly stable and sensitive biochemical sensors including both enzymatic and ISE sensors, which has obtained a record-breaking long-term stability of more than 100-hour continuous operation with negligible sensor degradation in sweat, and withstand complex matrix in wound. With the robust sensor development, we applied these sensors for HMI, wound management, and continuous prolonged daily activities monitoring. We also developed an ML pipeline that could reveal high accuracy and reliability for stress evaluation for the first time. We envision that by capturing a broader range of signals through more integrated biosensors, a more complete metabolic profile can be achieved for next-generation healthcare and human performance monitoring. Our multimodal sensors and machine intelligence pipeline could pave the way for numerous practical wearable applications such as intelligent healthcare and personalized medicine. In this final chapter, we bring up future directions for these architectures.

6.1 AI-generated e-skin

Human skin possesses outstanding mechanical properties, including flexibility, stretchability, toughness, along with multifunctional sensing abilities. However, there are many unsolved material challenges to replicating key properties in artificial skin[1]. AI has been proposed to optimize materials discovery and sensor designs to autonomously redesign new e-skin patches[1,2]. AI can be integrated into the materials design process in three phases (**Fig. 6-1**). The first phase involves model prediction and patch design based on functional requirements: size, weight, lifetime, cost, and other material specifications; the second phase entails computational modeling and experimental validation; and lastly, the improvement of current databases and model accuracies based on the results.

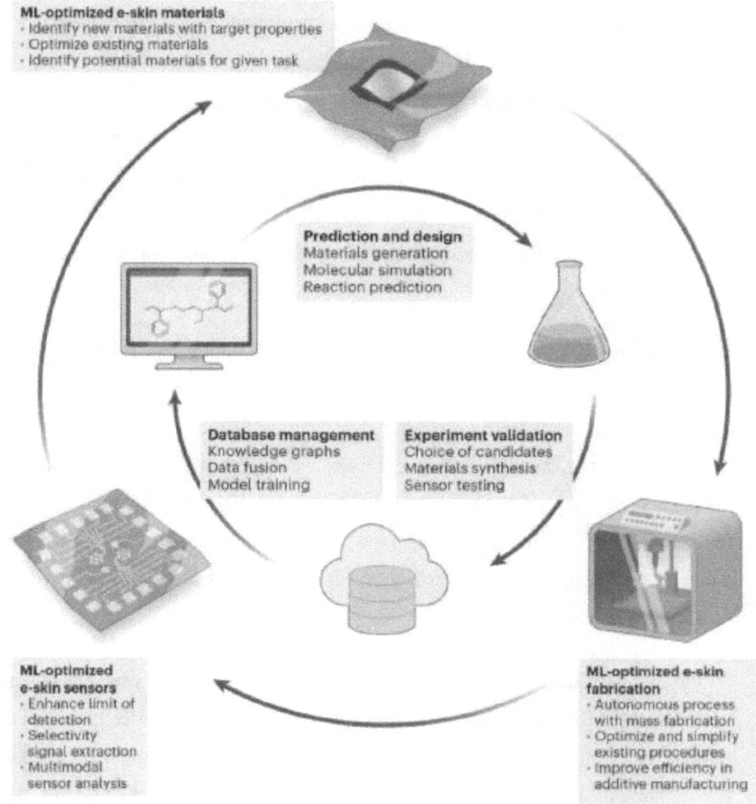

Figure 6-1. ML optimizations for e-skin designs. AI algorithms serve as an alternative pathway to optimize and explore materials synthesis, facilitate automatic mass fabrication and optimize current sensor limits.

6.1.1 Emerging materials and e-skin designs

The conventional selection of substrate materials typically involves natural materials such as cotton and silk, which are known for their biocompatibility, low-cost, and comfort.

However, natural materials have inherent limitations in stretchability and tunability. Material scientists and chemists consequently synthesize soft materials based on a combination of manual designs, drawing inspiration from nature, and leveraging previous material examples as references[3–5]. Some material design strategies include ultrathin tattoo-like substrates[6], applying serpentine interconnects[7], and using nature-inspired skin adhesion to realize high fiducial signal collection[8]. Meanwhile, these materials and designs require extra validation to characterize their properties, and many synthetic processes involve toxic precursors and require careful biocompatibility tests.

With a diverse availability of material candidates, designing or selecting a material with desired properties for a specified task is becoming increasingly challenging[9]. ML provides an attractive pathway to explore new materials and identify promising candidates with targeted properties, including alloy materials[10], NP synthesis[11], and electronic materials[12]. To date, a number of publicly available databases have been launched for simulating functional materials and recipes[1]. Moreover, ML can also be used to optimize and explore material synthesis, such as extracting text from scientific literature and giving synthesis protocol suggestions[13,14].

AI can help select and optimize fabrication methods based on material characteristics. Additionally, ML is can assist in quality control during mass fabrication, such as with jet printing of electronic circuits[15]. In addition to materials and fabrication methods, ML is also capable of optimizing e-skin designs. For example, a ML-based circuit designer has enabled transistor sizing adjustments using graph convolutional neural networks[16]. While conventional e-skin designs from planar designs typically do not conform to curvy surfaces[17], ML can guide structural designs of e-skins by finding kirigami designs for 3D shape-adaptive e-skins and pixelated planar elastomeric membranes more efficiently than mechanical simulations[18,19].

As most data from material experiments are discrete and noisy with high variance, it is necessary to preprocess the data through interpolating missing data and rebalancing biased training sets[20,21]. Additionally, many material science fields are not data-rich, and anthropogenic biases in the limited dataset may hinder model generalization[21]. This can be particularly true for collecting data about novel materials for human subjects. It is anticipated that a more standardized materials dataset and pipeline will speed up materials development and discovery[2].

6.1.2 Signal processing and augmented sensor performance

While traditional intuition-driven sensors are based on situation-specific experimental trials and time-consuming numerical simulations, ML algorithms can search for optimal sensor architectures as a function of required material properties with an accelerated and efficient prediction time[22,23]. In addition to conventional task-specific and labor-intensive signal processing, ML is capable of fast, robust data analysis to provide transferrable frameworks under different initial conditions. For example, ML can perform signal denoising[24], multi-source separation[25], artefact identification and elimination[26]. Two crucial guidelines for e-skin sensors are sensitivity and selectivity to the target biomarker. Indistinctive signal-to-noise ratios and overlapping detection between targets and interferents are two main bottlenecks for applying sensors for trace-level molecular detections in complex biomatrices. Substrates with similar structures to the target in biofluids could lead to confounding results. ML has been illustrated to improve the specificity and sensing limit of detection in multimodal sensing[27]. Many biochemical sensors involve enzymes that have a narrow working range, while AI algorithms could surpass signal saturation and calibrate non-linear sensors in a dynamic testing environment[28].

Motion artifacts are another major source for background noise in e-skins. While extensive analog and digital signal processing techniques have been applied to reduce artifacts and improve data quality[29,30], they typically involve manual circuit designs and simulations, which entail high costs and are not easily expandable to different scenarios. ML can be used for precise data acquisition by compensating noise and defects in wearable sensors[31]. In addition, data acquisition hardware can be fundamentally redesigned for optimal sensing with an intelligent platform[32,33]. The improved sensing capabilities as well as compact systems will fundamentally enhance sensor performance through iterative analysis of data-driven sensing outcomes[23].

6.2 AI-powered e-skin for human-machine interactions

HMIs enable the interaction between users and robotics, and have become crucial in remote robotic teleoperations. As the demand for precise and intuitive robotic control continues to grow, research has been turning its attention from conventional control theory towards a more immersive and interactive interfacing platform. The emerging AI-powered e-skins are creating new paradigms for robotic control and human commanded perception (**Fig. 6-2**)[34,35]. AI could quickly analyze multimodal data from e-skin patches and make autonomous decisions to manipulate robotics and provide human aid, which has already bridged the gap between human and machine interactions.

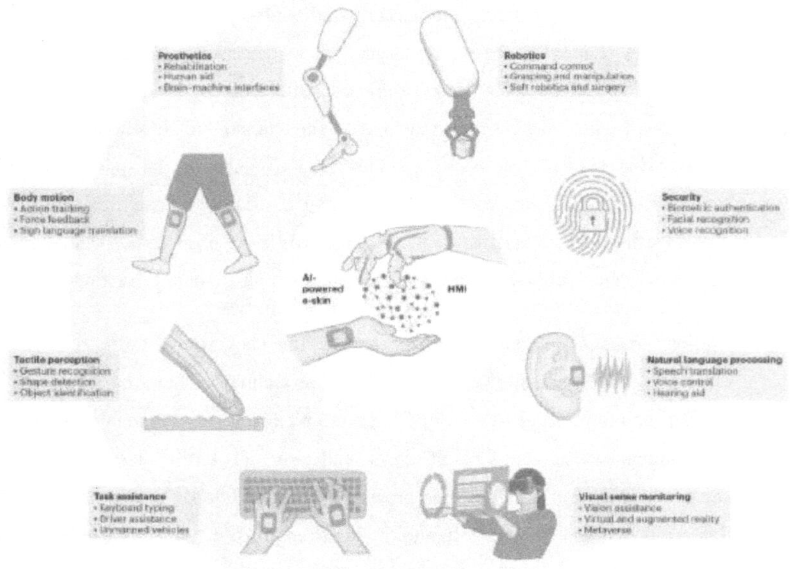

Figure 6-2. AI-powered e-skin for HMI. ML bridges the gap between humans and machines through task assistance, robotic control, virtual and augmented reality.

6.2.1 Tactile perception

Tactile perception decodes and transmits physical information to a computer system about hand movements, gestures, and force recognition[36]. The associated robotics can then accomplish tasks such as object grasping[37], shape detection[38], and object identification[39]. Haptic sensors are therefore widely adopted as a fundamental element for e-skin based HMI systems, which are usually built with arrays of strain and pressure sensors or electrophysiology electrodes such as surface EMG electrodes to capture complex hand movements[40,36,41,42], producing a large quantity of continuous data. Real-time haptic perception with the aid of AI has made tremendous progress in dynamic whole-body movements[42], gesture interpretation[43], tactile recognition[41,44], as well as object manipulation and detection[45].

6.2.2 Prosthetics and robotic feedback

Developing prostheses that rehabilitate motion for people with disabilities is a crucial goal in machine intelligence. Prosthetics typically involve a large sensing area with robotic feedback, where the e-skin extracts motion or audio data and ML algorithms analyze and control robotic operations accordingly. Strain and pressure sensors are fundamental components for actuators and grippers in robotics, enabling tactile feedback for enhanced functionality[41,46]. A variety of prosthetic solutions have been developed for different scenarios, including facial expressions[47], robotic control and feedback[38], translation of sign language into speech[48], personalized exoskeleton walking assistance[49], as well as providing steering and navigation assistance for people with impaired vision[50].

Smart robotic hands for prosthetics can also be applied for task assistance in healthy people. For example, a nanomesh-based e-skin integrated with meta-learning could assist rapid keyboard typing with a few-shot dataset[37]. Smart e-skin also has the potential for driving assistance by monitoring the driver's state and preventing sleep deprivation-related accidents[51], which provides an alternative solution for vehicle automation.

6.2.3 Hearing aid and natural language processing

Verbal communication with machines is another promising e-skin application that relies on AI, where a voice-user interface leveraging natural language processing is highly intuitive and convenient. Numerous studies have developed resonant acoustic sensors in e-skin for voice recognition[52], vocal fatigue quantification[53], and voice control of intelligent vehicles[54]. These sensors integrate resistive or piezoelectric membranes as sensing components[52,55,56], which converts human hearing range of around 20 Hz to 20 kHz. The customized frequency filtering can identify physical activities with different intrinsic frequency bands[55], or filter acoustic vibrations against human perspirations and background noise[57]. Voice sensors may also serve as a security device for biometric authentication[56].

6.2.4 Virtual and augmented reality

VR and augmented reality (AR) create a virtual environment where visual and auditory stimuli replicate sensations in the physical world[58]. E-skin provides an additional sensation of touch due to its unique skin interface[59]. For example, wireless actuators could be integrated in e-skins for programmed localized mechanical vibrations[58]. Such mechanical feedback can also form a closed-loop HMI system for motion capturing and vivid haptic feedback when interacting with virtual objects[60,61]. To further implement gesture controls for VR, a textile glove was developed with ML algorithms to classify hand patterns in various VR games[62]. AI could accelerate machine vision processing by utilizing a simple image sensor array matrix[63], empowering a high frame rate in VR visualizations. Additionally, some pioneering demonstrations have illustrated the potential of odor generators for olfactory VR applications[64].

6.3 AI-powered e-skin for healthcare and diagnostics

E-skin with arrays of integrated sensors can record the health profile of an individual in remote and community settings, detect aberrant physiology over time, and unveil health distributions at a population level. ML has aided diagnostics by identifying complex relationships between input physiological information and disease states[65–67]. There is a growing trend using AI-powered e-skins to address the growing demands in health monitoring and diagnosis (**Fig. 6-3**). Emerging AI has shown promising capabilities in approaching expert-level diagnosis, which could reduce the rate of misdiagnosis and create great clinical and market potential. For complex disease syndromes without established biomarkers, these ML algorithms could also facilitate our understanding in biomarker discovery, psychological predictions, and precision therapy.

6.3.1 Cardiovascular monitoring

Heart failure can worsen progressively over days while current telemedicine tools are not sufficient to detect acute exacerbations. AI-powered e-skins hold the promise of specialist-level diagnosis for cardiac contractile dysfunction or arrhythmias[68,69]. E-skins can integrate multiple modalities and facilitate the rapid evaluation of hemodynamic consequences of heart failure[70]. ML has been widely adapted for data analysis to extract cardiac parameters, such as blood pressure predictions[71,72] and left ventricular volume[73]. AI-based e-skin is anticipated to spot small and gradual cardiovascular changes over time and facilitate automatic diagnosis in a timely manner[70]. Such an approach will also alleviate the clinical load of physicians by reducing unnecessary hospital consultations.

6.3.2 Stress and mental health

Stress and mental health are significant problems for global health but their assessments rely heavily on subjective questionnaires. Pioneering studies for mental health predictions have been introduced including stress[74–76] and fatigue[77–79], but most studies still focus on commercial wearables such as watches which only monitor physical vital signs and are prone to motion artefacts. Several pioneering studies have demonstrated dynamic monitoring of the stress hormone cortisol using e-skin devices[80,81]. Next generation e-skins will combine physiological data with molecular signatures and perform multimodal data analysis[82]. By identifying previously unrecognized associations between health patterns and stress risk factors[83], smart multimodal e-skins with the aid of AI have the potential to model risk associations and unveil stress outcomes for mental health.

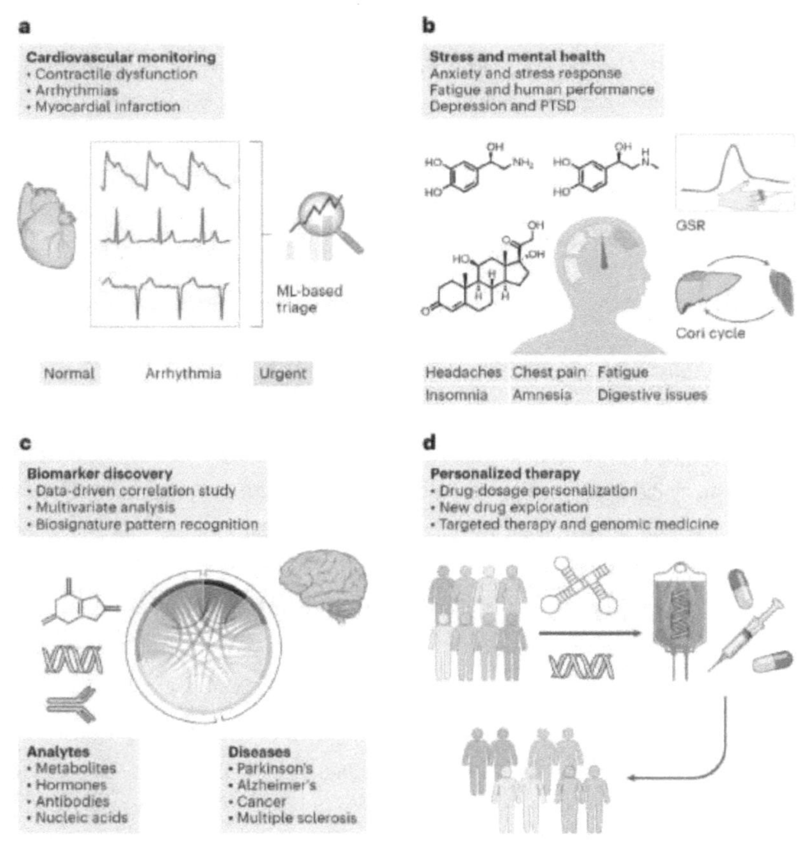

Figure 6-3. AI-powered e-skins for personalized healthcare and predictive disease diagnostics. a, Cardiovascular health can be investigated through continuous monitoring of one's cardiac activities (ECG, pulse waveforms, etc.) with e-skins. Integrating autonomous analysis through AI algorithms creates further potential for screening urgent conditions such as arrythmias. **b,** The application of AI-powered e-skin can extend to mental health which is a complex event that involves behavioral and physiological responses, metabolic changes, and fluctuations in a number of stress hormones. PTSD, post-traumatic stress disorder. **c,** Biomarker discovery through AI algorithms will further aid in finding new missing information potential links between measured sensor data and health status of individuals. **d,** Personalized therapy can be achieved by measuring individual's genetic and metabolic status using e-skins to develop highly targeted medicine for medical treatment.

6.3.3 Biomarker discovery

The development of AI is driving advances in both medical diagnosis and fundamental studies. Given the quantity of data in clinical studies, ML could be a transformative technology for data-driven biomarker discovery[84]. ML-based algorithms perform automatic data analysis for biomarker prediction, including skin disease[85], dysphagia[86], seizure[87], and COVID-19[88], where multiparametric monitoring based on multimodal e-skin platforms can reveal correlations between sensors and target outputs[89]. For diseases such as Parkinson's disease where no known effective biomarker is available, ML has the potential to unveil underlying correlations from the multi-dimensional data[90].

6.3.4 Personalized therapy

The development of drug and metabolic monitoring using e-skins has also aided in personalized therapy. AI-powered e-skins could benefit drug dosage personalization, where multimodal data coupled with ML models can be applied to evaluate pharmacokinetics and pharmacodynamics for personalized dosage[91,92]. Additionally, dynamic treatment of a disease affected by the individual's history and current course of action is well suited for the sequential decision-making used in reinforcement learning[93]. Prospective cohort studies involving physiological, metabolomic, environmental, and genomic data are anticipated to pave the way for the advancement of personalized therapy through the integration of AI-powered e-skin.

With the continued development and innovations in AI-powered e-skin, next generation e-skin is expected to aid prosthetics and the discovery of diseases, yet there remains several major bottlenecks including data acquisition and handling, data security, and data generalization. Data handling in both quantity and quality has become a challenge for model deployment. AI-driven data analytics are typically data-hungry, and training models with high prediction accuracy depends on large amounts of high-quality labeled data. Mature models such as decision trees and support vector machines demonstrate great accuracy and reproducibility and find extensive applications, yet their reliance on structured and manually labeled data poses high acquisition costs. In contrast, unsupervised learning unveils hidden patterns in unlabeled data, albeit with reduced accuracy and constrained applicability. Recent advanced models such as transformers have shown success in language processing and generation, but these models are of high complexity and require pre-training over big data sources using resource-intensive computing, with the underlying mechanisms still insufficiently understood. The time-continuous datastream from e-skin sensors carrying large amounts of unlabeled and heterogeneous data poses high demand for data processing and system integration. This necessitates a fast and cost-effective system for collecting and transmitting data to cloud-computing-based e-skins, while high-performance computing and storage units with low latency are required for in-situ applications[66]. Despite the growth in AI-driven e-skins, comprehensive regulatory frameworks addressing data accessibility, ownership, and security are yet to be fully established. This is crucial as public perception of data privacy risks can directly influence the adoptability of wearable devices, while user acceptance to disclose their medical information is uncertain at present[94]. While latest ML algorithms such as GPT-4 models have been reshaping the world, the success of large language model (LLMs) stems from the enormous amount of publicly available Internet data, which may not apply to the privately restricted medical datasets. Accessing regulated medical records and data poses significant challenges as they are highly restricted and obtaining them entails stringent protocols and privacy considerations[95], and data differences may potentially result in divergence from training accuracy. The FDA has recently updated its guidelines for handling sensitive medical data after announcing a new Office of Digital Transformation in 2021. Data generalization originating from built-in bias is another issue that could harm marginalized groups of people, which warrants special consideration for adopting ML models in medical practice. AI models can often make mistakes, but it is unknown who or what will be held responsible for controversial behaviors and outcomes of AI systems. Although models will become more powerful and capable over time, to what extent people can trust the ML predictions is still unknown[95]. The ability of fact-checking versus proof-reading may be beyond the expertise of users without clinical expertise[96]. Studies on interpretation and explanation of AI may be a possible solution[97].

From an e-skin perspective, another challenge is collecting high-quality biochemical data. Dealing with enormous amounts of rapidly fluctuating unlabeled data during continuous health monitoring may have adverse effects on model learning. Minimizing motion-induced artifacts from both the human and robotic bodies have required a strong interface and wearing comfort, and therefore poses need for strict materials properties, including biocompatibility, permeability, durability, mechanical strength, and conformability[98,99]. Biocompatible and non-toxic materials with strong, breathable and reversible skin adhesion are highly desirable for prolonged daily wearing, where the durability lifetime may depend on the specific use case[100]. Data accuracy can be improved by implementing multimodal sensing using one integrated platform to reduce defects from a single sensor[101]. Moreover, despite their high correlation with multiple potential diseases[102], many biochemical sensors struggle with low sensor stability, the necessity for frequent calibrations, and difficulty in detecting low-concentration biomarkers, which cannot provide as high-quality data as electrophysiological ones. Additionally, sensor embodiment and system integration is of concern when considering power sources, sensor arrays, signal processing and wireless data transmission[99]. Most integrated e-skins are powered through bulky rechargeable lithium-ion batteries; however, more research into wireless and low-power energy harvesting and storage is needed to develop fully flexible and sustainable e-skins[103,104]. These challenges have opened the door to exciting new opportunities in improving electronic sensors, optimizing patch designs, integrating cloud storage, protecting data privacy[105], and interpreting model accuracy[97]. The interdisciplinary collaborations among materials scientists, chemists, engineers, physicians, and data scientists are crucial to realize the full potential of the e-skin. The emergence of AI-powered e-skin marks a new era in the field of robotics and healthcare and is envisioned to transform the way human interacts with robotics and revolutionize medical diagnostics.